ENVIRONMENT AND SUSTAINABLE DEVELOPMENT IN INDIA

ENVIRONMENT AND SUSTAINABLE DEVELOPMENT IN INDIA

Edited by

RAJ KUMAR SEN
AMIT MUKHERJEE
PRAN KRISHNA PAL

Published on behalf of
CENTRE FOR STUDIES ON
ENVIRONMENT AND SUSTAINABLE DEVELOPMENT
RABINDRA BHARATI UNIVERSITY

DEEP & DEEP PUBLICATIONS PVT. LTD.
F-159, Rajouri Garden, New Delhi - 110027

ENVIRONMENT AND SUSTAINABLE DEVELOPMENT IN INDIA

ISBN 978-81-8450-280-0

Typeset by THE LASER PRINTERS, 8/15, 3rd Floor, Subhash Nagar, New Delhi-110027.

Printed in India at MAYUR ENTERPRISES,
WZ Plot No. 3, Gujjar Market, Tihar Village, New Delhi - 110 018

Published by DEEP & DEEP PUBLICATIONS PVT. LTD.,
F-159, Rajouri Garden, New Delhi-110027. Phones: 25435369, 25440916.
E-mail: ddpbooks@yahoo.co.in • ddpubs@gmail.com
Sales Showroom: 2/13, Ansari Road, Daryaganj, New Delhi-110002
Phone/Fax: 23245122

Professor Karunasindhu Das
Vice-Chancellor

RABINDRA BHARATI UNIVERSITY
Jorasanko Campus
6/4 Dwarakanath Tagore Lane,
Kolkata-700 007 Ph. 2269-1328
Emerald Bower Campus
56A Barrackpore Trunk Road,
Kolkata-700 050 Ph. 2556-8019
Fax: 91-(033)-2556-8079,
E-mail:rbreg@cal3.vsnl.net.in
Resi.Ph. 2642 7251

FOREWORD

Centre for Studies in Environment and Sustainable Development of Rabindra Bharati University has already been able to put forward its claim as a seat of serious studies and research in social science. In recent past, it organized seminars of utmost social relevance. It is heartening to note that the Centre now strides in publishing the papers read out in the seminar entitled Environment and Sustainable Development in India. I wish this venture all success and urge upon the Directorate for exploring new areas of relevant academic disciplines in future.

17 September, 2009

Karunasindhu Das

Contents

SECTION III
ENVIRONMENT AND SUSTAINABLE DEVELOPMENT: SPECIFIC ISSUES

SECTION IV
SUSTAINABLE DEVELOPMENT: SECTORAL ISSUES

LIST OF CONTRIBUTORS

1. **Dr. Raj Kumar Sen,** Professor of Economics and Director, CSESD, Rabindra Bharati University.
2. **Dr. Amit Mukherjee,** Finance Officer and Joint Director, CSESD, Rabindra Bharati University.
3. **Dr. Pran Krishna Pal,** Reader in Economics and Co-Director, CSESD, Rabindra Bharati University.
4. **Dr. V.R. Panchamukhi,** Managing Editor, Indian Economic Journal and Former Chairman, ICSSR.
5. **Dr. S.D. Chamola,** Former Professor of Economics, CCS Haryana Agricultural University, Hisar.
6. **Dr. Ratan Lal Basu,** Former Teacher-in-Charge, Bhairab Ganguli College, Kolkata.
7. **Dr. M. Anisur Rahman,** Former Vice-Chancellor, Rajsahi University, Bangladesh.
8. **Dr. Kausik Gupta,** Professor & Head, Economics Dept., Rabindra Bharati University.
9. **Ms. Amrita Pal,** Dept. of Economics, B.B. College, Asansol.
10. **Dr. Jayanta Sinha,** Dept. of Zoology, B.B. College, Asansol.
11. **Dr. Pranab Nag,** Dept. of Commerce, Nabadwip Vidyasagar College, Nabadwip.
12. **Sri Anup Kumar Saha,** Dept. of Economics, Nabadwip Vidyasagar College, Nabadwip.
13. **Dr. G. Sathis Kumar,** Research Scholar in Economics, Gandhigram Rural University, Gandhigram.

14. **Dr. S. Ramaswamy,** Professor and Head, Dept. of Economics, Gandhigram Rural University, Gandhigram.

15. **Dr. Sheela Datta Ghatak,** Dept. of Zoology, P.D. Women's College, Jalpaiguri.

16. **Dr. Asok Dasgupta,** Dept. of Anthropology, North Bengal University, West Bengal.

17. **Sri Anirudha Ojha,** Research Fellow, Utkal University, Bhubaneswar.

18. **Ms. Mala Bhattacharjee,** Dept. of Economics, Raja Peary Mohan College, Uttarpara.

19. **Ms. Rajrupa Mitra,** Dept. of Economics, Raja Peary Mohan College, Uttarpara.

20. **Dr. Somnath Hazra,** Programme Manager, RCDC, Bhubaneswar.

21. **Prof. Binayak Rath,** Vice-Chancellor, Utkal University, Bhubaneswar.

22. **Sri Har Govind Prakash,** Research Scholar in Economics, Allahabad University.

23. **Sri Anath Bandhu Mukherjee** (Retd.), Economics Dept., S.R.F. College, Beldanga.

24. **Dr. P.S. Tripathi,** Dept. of Economics, D.B.S. College, Kanpur.

25. **Ms. Shweta Dikshit,** Research Scholar, D.B.S. College, Kanpur.

26. **Dr. Subrata Kumar Ray,** Dept. of Economics, Sabang Sajanikanta Mahavidyalaya, West Bengal.

27. **Sk. Nazrul Hussain,** West Bengal.

28. **Ms. Rifat Mumtaz,** National Centre for Advocacy Studies, Pune.

Environment and Sustainable Development in India: An Overview

Raj Kumar Sen

The modern western economic theories never considered environment seriously and considered it, if at all, as a free good like air, light and water. The problem of environmental pollution was at best considered as external diseconomies in some cases. Only in recent years and in the decade of the seventies of the last century in particular, environmental concerns started to figure in a prominent manner in economic thinking as a consequence of the occurrence of acid rain in several parts of Europe. Ultimately the 1972 summit at Stockholm articulated rigorous thinking about degradation of environment due to current human activities and since then environmental studies have taken an important place among the thinkers as well as policy-makers of both developed and developing countries. Gradually the related concepts like ecology, biodiversity, conservation of natural resources, sustainable development and others started to emerge and in economics also a new branch called environmental economics has now come to stay.

However, in pursuing such studies we often face a number of problems. Thus environment is not a homogeneous entity and varies not only from one country to another, but also between the different regions of the same country. Naturally, it is not possible to adopt same or similar strategies to combat environmental problems at different places. Moreover, as environment primarily deals with physical resources and a subject of material sciences, it is often difficult to frame economic policies on this subject which is basically multi-disciplinary in nature. Further, though there is an acceptable definition of sustainable development as per the Bruntland Commission, yet under the current development paradigm of neo-liberal thinking, it is not very easy to implement in actual practice the operative part of it. An attempt is made in this volume to understand the problems related with environment and sustainable development and especially those which are particularly relevant for India. As many as nineteen articles mostly selected from invited and contributed ones presented in a recent National Seminar on the same theme are included in this title. These are distributed over four sections considering the focus of these articles and these are briefly introduced in the following paragraphs. However, for the benefit of readers, all articles are accompanied by an abstract. Also in the appendix of this volume there is the proceedings of this Seminar from which one may get an idea of these and other articles discussed there.

After this overview, the section 1 opens on the theme **Environment and Sustainable Development: Concepts and Ideas** containing altogether four articles. The first article is on 'Globalization and Sustainabilities—Issues and Challenges' **(V.R. Panchamukhi,** the Chief Guest of the Seminar) where the author dwelt upon a number of issues and concluded that: (a) There is an urgent need for a new paradigm of world order as the current process of globalization is reaching the brink of collapse; (b) The concept of sustainability is much wider than mere environmental sustainability though sustainable use of water resources is especially important; (c) Sustainable process of development should be practised in a holistic framework; and (d) In this context ancient Indian thoughts on environment can provide many solutions even to the contemporary world. This last issue has been dealt elaborately in the keynote paper **(S.D. Chamola)** for this section in the context of present unsustainable development strategy. A specific discussion of environment and ecology as given in Kautilya's *Arthasastra* is the subject matter of the next paper by **Ratan Lal Basu.** This section comes to an end with the paper on environment and sustainable development **(M. Anisur Rahman**, Special Guest from Bangladesh) focusing on the need to maintain biological diversity, giving an international flavour to this national seminar.

A set of five papers are included in the next section on **Sustainable Management of Natural Resources and Environmental Governance. Kausik Gupta** (Keynote paper for this section) with **Sanchita Som** took up the issue of sustainable management of renewable resources in the Indian context. Forest and fishery management are the two issues focused specially in their paper. Next **Amrita Pal and Jayanta Sinha** discussed in their paper on Environment Performance Index which led to Environment Sustainability Index. They also discussed their relationship with several macro-economic indicators. Micro and macro-aspects of environmental accounting and their role in developing an appropriate project appraisal model as a key to sustainable development has been discussed in the next paper **(Pranab Nag and Anup Kumar Saha).** The three-tier participatory Global Environmental Governance Mechanism for providing a sustainable society is the theme of the joint paper by **G. Sathis Kumar and S. Ramaswamy.** Finally with the paper by **Sheela Datta Ghatak** emphasizing the need for community participation in the context of the relationship between environment, health and education, this section comes to an end.

The next section is confined to the discussion of some specific issues and there are four papers in it. **Asok Dasgupta** has taken up the case of the Rajbansi community to discuss the relevance of the indigenous peoples. **Anirudha Ojha** has concentrated in his paper on the tribal belt of Orissa to discuss the poverty of the people in this region and their property rights and management problems of the common property land resources. Ecotourism as an environmentally sustainable development process is the theme of the next paper by **Mala Bhattacharjee and Rajrupa Mitra** who have discussed it in the context of Sonamukhi forest of Bankura, West Bengal. The last paper in this section is by **Somnath Hazra and Raj Kumar Sen,** who with some success stories have discussed the ecological approach to remove chronic poverty in India for the sake of sustainable development.

The last section of this volume has dealt with the sectoral issues for sustainable development and there are as many as six papers included under it. **Binayak Rath** (Special Guest) has discussed the Indian mining sector under globalization and took up a sustainable development approach in terms of environment impact assessment and corporate social responsibility. **Har Govind**

Prakash has taken up the problem of slums in urban India in the context of preservation of environment and sustainable development. In a similar vein **Anath Bandhu Mukherjee** has discussed the problem of the special economic zones and the pollution created by them. A new area—the role of banking sector for sustainable development and environment protection, has been elaborated by **P.S. Tripathi and Shweta Dikshit** in their joint paper. The pollution created by the Kolaghat Thermal Plant in West Bengal and its effects on agricultural production are discussed in the next paper by **Subrata Kumar Ray and Sk. Nazrul Hussan.** The last paper of this section as well as of the volume is also on a similar topic viz. the pollution impacts of the Indian sponge iron industry by **Rifat Mumtaz.**

This publication is one of the important activities of the newly established CSESD (2008) of the Rabindra Bharati University. We are thankful to the University authorities for providing physical and financial inputs to organize its activities. Special thanks are due to our Vice-Chancellor Professor Karunasindhu Das who have inspired us by writing the encouraging Foreword to this volume. We are extremely thankful to the invited and contributing authors for submitting their papers in the form of articles without whose valued cooperation this volume could not see the light of the day. The directors, advisers and working committee members of the CSESD and the Centre for Higher Education Management (CHEM), RBU (who lent a collaborative hand to the CSESD in organizing the said National Seminar) deserve the heartfelt thanks from the editors for their cooperation and support. We are also thankful to the Environment Department, Govt. of West Bengal, ICSSR-ERC, UBI (SBM-E) and an anonymous funding agency for their generous support to this Seminar. Last but not the least, we are extremely thankful to our good old friend, Shri G.S. Bhatia of Deep & Deep Publications Pvt. Ltd. for publishing this volume in such a meticulous manner. We sincerely hope that this volume will be useful to the researchers, teachers, senior students, policy-makers and to the general readers interested in environment and sustainable development and it will be both light and fruit bearing by expanding the frontiers of this very important theme.

SECTION I

ENVIRONMENT AND SUSTAINABLE DEVELOPMENT : CONCEPTS AND IDEAS

Globalization and Sustainabilities: Issues and Challenges

V.R. Panchamukhi

In the recent years, globalization has become an important feature of the development strategy in most parts of the world. It is expected that globalization would help the countries in improving efficiency of resource allocation and also in realizing higher growth in GDP. However, in practice, adoption of the strategy of globalization has not only belied these expectations but also has generated many unprecedented hurdles in the task of realizing optimum welfare for the peoples at large. In this paper, an attempt has been made to examine the *impact* of the process of globalization on the sustainability of the development process. It is argued that globalization, as practiced to day, has implied iniquitous distribution of the benefits of globalization and that there is need for *a fair globalization*. It is also argued that the world has witnessed a series of *crisis situations*, over the past few decades and that the world has not learnt *suitable lessons on the corrective measures required to avert the recurrence of crisis situations*. Urgency of evolving a new *financial architecture* has been underscored. The paper raises the question of sustainability of the present wave of globalization and puts forward the concept of *holistic development*, which implies a harmonious blend of three dimensions of the process of development, viz. growth of GDP, social aspects of development and preservation of cultural values and lifestyles. As an illustration, the paper puts forward an outline of a *new world order*, in regard to the systems of education and use of water resources, aimed at realizing sustainability. The paper also gives a brief outline of the ancient Indian thoughts on the concepts, such as, development, globalization, sustainability, environment, and values of life.

INTRODUCTION

I would like to compliment CSESD, in particular, its Director, my good friend, Professor Raj Kumar Sen for having taken the initiative to organise such a timely and unique National Seminar. I am also thankful to CSESD for inviting me as the Chief Guest. I would also like to express my appreciation and gratitude to Professor Karunasindhu Das, Vice-Chancellor, Rabindra Bharati University, a good scholar in Indology, for having spared his time to be present here and inaugurate the Seminar. I have chosen the theme entitled, "Globalization and Sustainabilities—Issues and Challenges", for sharing some of my thoughts with this august assembly. I propose to raise some broad Issues of analytical and practical relevance, with a view to *provoking* some critical debate on this important subject. In my address, I propose to deal with the following issues:

- Reflections on the Broad Concept of "Sustainability";
- What are the Contours of the concept of "Environment"!;
- Some Questions on "Sustainability";
- Is the present wave of globalization sustainable?;
- Some select issues, such as, a New paradigm for sustainable Use of Water and UN Decade of Education for Sustainable Development;
- Episodes and causes of Crisis Situations;

- Need for Holistic Development; and
- Some insight into the Indigenous "Thoughts on Sustainable Development."

CONCEPTS OF SUSTAINABILITY

The concept of Sustainable Development has become a common theme in the debates on development strategies, ever since the famous Brundtland Commission introduced this concept in its celebrated Report in the mid-1980's. Even though the concept of sustainability was introduced in W.W. Rostow's thesis of *Take-off into self-sustained growth,* it should be noted that it was in a different context than the one in which, the Brundtland Commission developed the notion of *sustainable development.* Brundtland Commission Report defined *sustainable development* as that development which "meets the needs of the present without compromising the ability of future generations to meet their own needs". The concept was more in the context of the impact of the development process on environment and *vice versa.* However, the notion of sustainable development, implied in the Rostow's theory was pertaining to the structural features of the economy, referring to the parameters, such as, savings rate, investment rate, structural changes in production and trade and productivity in resource use. The question here was the following: Would the economy be able to continue to maintain its growth momentum, on the basis of its own structural strengths? In my view, the concept of sustainability should be considered in its wider framework for evaluating the development process and also for evolving suitable development strategies for the future.

The literature on sustainable development has classified the concept of *sustainability* into four categories, viz: *Environmental Sustainability, Economic Sustainability, Socio-Political Sustainability and Cultural Sustainability.*

Environmental Sustainability is to be understood in the framework of sustainable development, introduced in the Brundtland Commission Report, referring to the interface between growth and environmental impacts. *Economic Sustainability* should be considered in the wider context of structural features of the economy, as propounded in the Theory of W.W. Rostow. The notion *of Socio-Political Sustainability,* would refer to the question as to whether the desirable social order and the political system, would be sustained in the future. Lastly, the issue of *Cultural Sustainability* would refer to the question as to whether the Cultural values and lifestyle etc. of the society would be sustained in the context of the fast changing global order.

It is useful to discuss the implications of the present wave of globalization and those of the manner in which globalization is being enforced in the world, for all these categories of sustainabilities. It should be noted that the literature on environmental sustainability has conceived different gradations of sustainability, such as, *Weak Sustainability, Strong Sustainability and Deep Ecology.*

Deep Ecology develops an *ecological wisdom* by focusing on *deep experience, deep questioning and deep commitment*. This concept inducts *metaphysical dimension* to the debate of *sustainability. Deep Ecology* perceives *humans as* part of the *Biotic Community, consisting of humans, non-humans and nature.* The contours of *Deep Ecology* attempt to integrate *Ecological Consciousness* with what one may call as *Spiritual Consciousness.* The analysis and prescriptions of *Deep Ecology,* remind us about the framework and prescriptions of Indian Philosophy, in particular, those of *Isavasya Upanishad* and of the *Holistic Approach of Vedanta.*

It is often feared that the nature of globalization as is taking place to day, would imply many negative effects on the momentum of growth in many developing countries. It is also argued that globalization is leading to increase in inequalities and collapse of cultural identities of the individual societies and nations. In short, the present form of globalization is threatening to destroy all the four forms of sustainabilities stated above. We would leave this issue as a Thesis on which further research and debate are called for.

MULTIPLE DIMENSIONS OF SUSTAINABILITY

It is worth-mentioning that the United Nations Division on Sustainable Development includes, among others, the following disciplines in the framework of the contours of sustainable development: Agriculture, Atmosphere, Biodiversity, Biotechnology, Climate Change, Consumption and Production Patterns, Demographics, Desertification and Drought, Disaster Reduction and Management, Education and Awareness, Energy, Ecology, Land management, Poverty, Sanitation, Waste and Water. Thus, a process of development would be, in principle, unsustainable, whenever there is some *mismatch among* the pace and pattern of the dynamics of the different parts of the system, identified above. Development Crisis is a manifestation of the phases of such mismatches. It is the responsibility of the academics and practitioners, to anticipate the impending episodes of such mismatches and identify necessary corrective measures.

CONTOURS OF ENVIRONMENT

Before proceeding to raise some specific issues for further discussion, let me present some reflections on the contours of *Environment*. Normally, *Environment* is conceived to be consisting of *land, water, fire, air and space (Prithivi, Ap, Tejas, Vayu and Akasha).* The modern theory and practice of environment talk about pollution of soil, pollution of water, pollution of air, pollution of space, etc. However, ancient Indian wisdom considers the contours of environment and its pollution in a much wider setting. The ancient Indian wisdom considers the following entities also, as the factors, defining the Contours of *Environment: Time (kaala), Direction (Dik), Conscience (Atma) and Mind or Thoughts (Manas).*

We rarely recognize that *Time* is one of the precious resources of mankind. Misuse of time, inefficient use of *Time,* time-overruns in project implementation, absence of punctuality in the public functions, etc. would imply pollution of the *Time-environment.* Of course, as per the astrological calculations, some periods of time, such as, *Rahu Kala, and Gulika Kala* are proclaimed as *inauspicious* for good actions. Even if we do not adhere to this approach for defining pollution in the use of time-resource, the concepts of *time and its pollution* would be relevant in our activities, as mentioned in the previous paragraph.

In the same way, *Direction* is also a resource and decent use of it could generate most unexpected good results. Again, in our ancient wisdom, we have the Theory and Practice of *Vastu Sastra,* which combines the spiritual assessment of direction with its physical effects. A factory or a house built with wrong use of directions could imply inadequate flow of fresh air and thereby generate health problems for the inmates. Hence pollution *of direction as a resource* needs to be avoided. *Atma* signifies *Conscience*. Obviously, any action done against the dictates of *conscience* would distort a value based social order. The environment of corruption, observed to day, clearly indicates the pollution of the *conscience as a resource*.

If *Mind* is polluted or if it nourishes perversions in the thought processes, then, every thing else would become polluted. *Management of Mind* is beyond the scope of the received theories and practices *of Sustainable Development*. We have to take recourse to the perceptions of the *Vedantic wisdom of India*, for evolving a framework for the management of Mind so that *pollution of Mind* could be avoided.

The brief digression given above may appear to take us to the domain of philosophy, much away from the present day conception of economics. However, I would like to assert that social science, which is totally disconnected with a *holistic philosophy of life*, implying a blend of *materialism and spiritualism*, would not serve the purpose of giving to us a framework of analysis aimed at the promotion of true welfare of mankind.

Let me now revert to the mainstream discussion of the subject.

Current Process of Globalizations: Some Reflections

It is common knowledge that in the recent years, *Globalization, Liberalisation and Privatisation* have become common *Mantras* in the context of Economic Reforms. However, there is very little attention given to the basic issue as to how the manner in which the new strategies are being implemented, would adversely affect the existing systems and hence raise the questions of sustainability of the development process itself. Let me briefly mention some of the anomalies that are emerging in the recent times.

- There is a growing divide between the emerging formal global economy and the expansion of an informal local economy. The latter is getting increasingly marginalised;
- The benefits of globalization have been unequally distributed, both between and within countries; Growth divergences have widened between the countries and the gap between the rich and the poor has also expanded;
- There is imbalance in the global Rules and Institutions—

Economic and Social; Global rules favour the rich to a greater extent than the poor; Volatility in capital markets has adverse effects on the poor;

- Trade in Manufactures has been liberalised while trade in agriculture remains protected, particularly, in the developed country markets;
- Asymmetry in regard to access to technology, information and knowledge persists;
- Structural Changes are taking place without adequate Economic and Social Provisions for Adjustments; this has resulted in increase of uncertainties and insecurities, in particular, for those whose "capabilities" remain static.

Some Candid Views on the Present Process of Globalization

Let me place on record, for facilitating an informed debate, some of the critical observations made by some experts on the present process of globalization, which have implications for the sustainability of the process itself:

- Lael Brainard of University of California, San Diego, writes: *"For any one in Latin America or Asia, globalization simply equals Americanization. Many worry that Globalization is a grand American Conspiracy—a relentless quest to subordinate the planet to Mickey Mouce, Madonna and McDonalds—or for today's hip grads, Laura Croft, and the Lakers ...Globalization equals marketization"*;
- Doni Rodrik wrote, *"Globalization is exposing social fissures between those with the education and skills, and mobility to flourish in an unfettered world market—the apparent "winners"—and those without....The result is severe tension between the market and broad sectors of society, with Governments caught in the middle;"*
- *Globalization is the realization of the Vision set forward by the "Trilateral Commission," set-up in the 1970's and headed by David Rockefeller (Chairman, Manhattan Bank) to move quietly to establish a "borderless" world in which the transnational corporations would be free of interference of the local states, so they could compete effectively in the "new world order;"*

- *Current Globalization* can be described as the phenomenon of *sacrifice of many* for the *benefit of select powerful few;* In contrast to this, the ancient Indian approach to *Globalization* implies *sacrifice of a few* for the *benefit of many;* and
- *Vasudhaiva Kutumbakam, Sarvey Janah Sukhino Bhavantu, Isavasyamidam Sarvam, etc.* characterise the ancient Indian conception of *Globalization.*

SOME QUESTIONS ON SUSTAINABILITY

Below is given an inventory of issues and questions that deserve attention in the context of the debate on *Sustainabilities* in the wider context. I have no pretension of dealing with all of them in my lecture. However, it is my belief that a comprehensive list would be useful for encouraging further debate and research on the subject. Below is given a list:

- Is the present wave of *Globalization,* as practiced to day, sustainable?
- Is the present *Paradigm of Development,* with market alone, in the driver's seat, sustainable?
- Is the present *Non-system of volatile Capital flows* sustainable?
- Is the present model of focus on *Growth Rate alone,* without considering the other dimensions of *Development* sustainable?
- Is the present pattern of Demand and Supply of Water Resources sustainable, in India and globally?
- Can the present pace and pattern of population growth and that of food production, sustainable in India and in the globe, with the existing inequalities in the purchasing powers?
- Is the present pattern of *land use in Indian Agriculture* sustainable in the medium and long-term?
- Is the pattern of trade liberalization and its inverse relationship with environment and *Food availability, sustainable?*
- Is the present system of *exclusiveness* in regard to the benefits of development and also in regard to the education and health facilities sustainable?

- Are the present pace of urbanization and speed of infrastructural development, sustainable?
- If the speed of Consumption of renewable resources is more than nature's ability to replenish, is this sustainable? This is so in many parts of the globe.
- Governance for sustainability has its origins in holistic awareness and competence, benign empowerment, social equality, and responsible values, visions, and actions. Is there any attempt to evolve a suitable model of governance for sustainability?
- Does the democratic form of political system facilitate enforcement of the rigours of sustainable development, without *reforming the citizen?*
- Do the stipulations of sustainable development conflict with the compulsions of poverty eradication and job creations?
- Has the gravity of the issues of climate change, been adequately responded to, by the growth of *ecological economics and adoption of suitable practical measures?* Is there any impact of all this on any paradigm shifts in technological choices, lifestyles and the agricultural practices towards more sustainable systems?

The list of concerns on the issue of sustainability could be expanded further. But it suffices to assert that the debate on *globalization and sustainability* should go much beyond the contours of mere *environmental sustainability.*

IS THE PRESENT WAVE OF GLOBALIZATION SUSTAINABLE?

R. Robertson categorizes, Three distinct Global Waves of Interconnectedness, which may be termed as "Globalization", as follows:

- The first Wave from the period 1500 A.D. to 1800, largely driven by urge for access to more and more of resources;
- The Second Wave from 1800 A.D. to 1945,—end of second world war, with the process of industrial revolution at the driver's seat;
- The Third Wave from 1945 to present times, driven by

what may be called as "Corporatism", American "Globalism" and "New Economy".

Struggle for Survival, Urge for Security and Well-being are the most common factors driving all these processes of globalization. Types of strategies adopted for ensuring survival, security and well-being, during all these waves of globalization, have been, in general, the following: Migration, Conquest, Commerce and Technological Innovation. It should be noted, some of these driving factors are present even in the most recent (Third) wave of globalization. In addition to the above, the "American Globalism" (after the decay of the "Soviet Globalism") is the dominant framework of the recent Wave.

Ironically, the very factors which generated Waves of Globalization, were responsible for creating the forces for their Collapse. Persistence of Hegemonic Tendencies, Exclusiveness in Development Processes, Increase in Inequalities, Accentuation of Divide between the traditional sectors and the modern sectors, as also continued marginalization of the traditional sectors, persistence of poverty and unemployment, are, some of the factors causing frustration of "security and well-being", of "some", leading to systemic collapses.

NEED FOR A FAIR GLOBALIZATION

The recent wave of globalization is evolving in such a manner that it has raised some basic doubts as to whether it is *fair and sustainable*. In this context, the United Nations had set-up a Commission entitled, the World Commission on the Social Dimensions of Globalization. The Report of this Commission, entitled "A Fair Globalization—Creating Opportunities for All", has advocated the need for a fair globalization, with a view to making globalization a sustainable one. It is observed that the present wave of liberalization, privatisation and globalization has the danger of creating a highly iniquitous world order. Many concerns of mankind, such as acute poverty and destitution, widening gap between the rich and the poor, growing unemployment, pattern of exclusive development, inadequate access to health facilities, drinking water and education, etc. are neglected. It is in this context, it is advocated that the Social dimensions of globalization deserve immediate attention. This Report also opines that launching of *Millennium Development Goals* by the UN is a clear indication of the inadequacies of the market system and globalization, for evolving a fair and sustainable global order.

SPECIFIC EPISODES OF CRISIS SITUATIONS

Let me present a brief profile of the series of crisis situations that have come to the fore in the recent decades, in the world economic system. The frequency and the intensity of these crisis situations clearly raise doubts about the sustainability of the present processes of globalization and structural changes. A brief reflection on the recurrence of the crisis situations also raises the question as to whether the world has learnt proper lessons from the crisis situations from time to time and whether the proper corrective measures have been identified and put into place. I do not propose to go into the details of these issues. But my purpose is to draw the attention of the academic and policy worlds about the series of crisis situations and urge them to critically examine the causes and effects of these crisis situations and to derive suggestions for a suitable New Paradigm of development in the world. I present below the list of the crisis situations in a synoptic manner and I do not intend to give more details of each of the situations. It is for the researchers to look into each of these situations and conduct intensive studies on them:

1980's: Latin American Debt Crisis, beginning in Mexico; African economic crisis with starvation despite pursuit of structural adjustment programmes;

1989-91: United States Savings and Loans Crisis;

1990's: Collapse of the Japanese Asset Price Bubble, Consequent stagnation of the Japanese economy, with its effect on others;

1991: Crisis of Balance of Payments, and Debt servicing in India;

1992-93: Speculative Attacks on currencies in the European Exchange Rate Mechanism;

1994-95: 1994 economic crisis in Mexico; Speculative Attack and default on Mexico's Debt;

1997-98: Asian Financial Crisis; Devaluations and Banking Crisis across Asia and economic crisis with contagion effects;
1998: 1998 Russian Financial Crisis; Devaluation of the ruble and default on Russian debts;
2001-2: Argentine Economic Crisis (1999-2002); Breakdown of the Banking System;
2008: USA, Europe; Spread of the US Subprime Crisis;
2008: Financial Crisis leading to Economic Crisis, with the danger of leading to *Recession, throughout the World.*

This frequent recurrence of *Crisis situations* is a clear manifestation of the *unsustainabilities* inherent in the present *Systems.*

Common Causes for Most of the Crisis Situations

It is rather puzzling to note that there is a common thread of causal factors which have been repeatedly cropping up to generate the crisis situations, which manifest themselves in different forms. These causal factors could be listed as follows:

- Excessive Opening up of the economies, before *adequate capability building for competition;* To put it differently, No integrated approach, adopted for *opening and domestic capability building;*
- Faster globalization of the Capital Markets, leading to high volatility in Capital Flows and exchange rates; Inappropriate sequencing of the so-called Economic Reforms;
- Lack of Coordination among macro-economic policies in the world;
- Mismatch between the expansion of the Financial Sector and the Real Sector of the Economies;
- Asset-Liability Mismatch; Increasing Non-performing Assets in the banking sector; Huge Debt Burden of the Governments;
- Neglect of the *traditional sectors and the social aspects of the development process,* leading to increasing unemployment, poverty and income disparities; Absence of suitable Safety Nets, etc.

PARADIGM WITH FOCUS ON GROWTH ALONE IS NOT SUSTAINABLE

While discussing the sustainability of the present market driven paradigm of development, with focus on growth rate alone, it is worth-recalling the reflections presented in the UNDP's Human Development Report, produced in the mid-1980's. This Report urged the policy-makers to avoid five types of growth, which cause non-sustainabilities, as follows:

- Avoid Jobless Growth; A Growth Profile, which implies "loss of job opportunities" should be avoided;
- Avoid Ruthless Growth; Growth process which implies aggravation of inequalities is not desirable;
- Avoid Futureless Growth; A Growth Process which implies non-sustainability should be avoided;
- Avoid Voiceless Growth; A Growth Process, which does not enhance "empowerment" of the derived sections of the society is not desirable;
- Avoid Rootless Growth; If the Growth Profile implies destruction of the cultural foundations and traditional values of a society, then such a growth profile is not conducive to the enhancement of human welfare and hence it should be avoided.

NEED FOR HOLISTIC DEVELOPMENT

As a result of the discussions presented in the previous paragraphs, we put forward a proposition that the world should evolve a New paradigm of development, which may be termed as *Holistic Development*. The concept of *Holistic Development* is based on the thesis that there is in important distinction between *Growth and Development*. It was recognized in the literature on development economics of 1960's that development is much wider concept covering issues such as equity, social justice, empowerment and human development, while *Growth* is a narrower concept confined only to the phenomenon of growth in GDP. The latter is obviously not a full indicator of the welfare of mankind. It is in this context that I introduce the notion of *Holistic Development*, which incorporates the following dimensions:

1. Growth of GDP or that of per capita GDP;
2. Social aspects of development, which include concerns, such as, unemployment, poverty, inequality, education, health, social infrastructure, empowerment, gender inequality, etc.; and
3. Values and Cultural aspects of development.

My purpose is to bring home the point that the present emphasis on Growth alone, as a criterion for choosing and evaluating the development strategies is faulty in so far as the growth process neglects the other important dimensions of the development process, viz. *social aspects and cultural and values aspects.*

Select Questions Before Us

In the context of our discussion on Globalization and Sustainabilities, let me identify a few specific issues and questions, which deserve immediate attention for research and policy analysis. These are spelt out below:

- Is India insulated from the contagion effects of crisis and recessionary tendencies in the rest of the world, as it was during the earlier crisis periods?
- If India is *exposed* to a much larger extent than what it had been about a decade ago, what needs to be done to minimize the *adverse effects?*
- Can we open up new debates on issues, such as, *optimum globalization, new financial architecture, sequencing of liberalization, new development paradigm, inclusive growth, role of the state in an integrated world system, governance for sustainabilities, fair trade and fair globalization, etc.?*
- Should we open new debates on issues, such as, Urgency of a New Financial Architecture, Need for a New paradigm of Holistic Development, Establishment of a New global Economic Order, Search for Relevant *Indigenous Thoughts.*

I have no pretensions of answering all these questions in my brief address. However, in the following paragraphs, I propose to give some synoptic account of some of these issues.

It is obvious that during the present global financial and economic crisis, India is not insulated from the external shocks, to the same extent as it had been during many of the previous crisis situations, including the one in the late 1990's in the neighbouring Asian region. This is because, India has been intensively integrated with the world economic system, to a much greater extent than before. Since many fundamental issues have come to the fore, there is need for raising a debate on a new paradigm of development and a new world economic order.

Urgency of a New Financial Architecture

International Community has recognized the need for a New Financial Architecture (NFA). Unfortunately, there is very little progress in practice, in this regard. It is, however, useful to recall the features of the NFA, as identified in the literature on the subject.

Some desirable features of NFA are the following:

- Efficient Allocation of Capital; strengthening nexus between financial flows and development priorities;
- Restoring the earlier system of "official capital flows" aimed at promotion of development;
- 'Managing' and 'regulating' free capital mobility;
- Providing financial safety nets to the vulnerable sections of the society;
- Addressing the issue of information asymmetries;
- Preventing "herding" in the financial markets;
- Improving trade-off between financial liberalization and financial stability;
- Preventing financial crisis and minimizing their adverse effects, when they occur; and
- Ensuring global stability and promoting economic growth with social justice.

The debate on reforms in the International Financial and Monetary System has been going on for a considerable period in the past, in fact, since the collapse of the fixed exchange rate regime, in the early 1970's. It may be recalled that the IMF had set-up an expert group, called, Committee of Twenty, which had, in its Report, cautioned the world leaders about the non-sustainability of the volatile exchange rate system and also about the adverse effects of the unregulated capital market systems. The Non-aligned Movement had set-up an expert group in the late 1980's, to review the world monetary order and suggest remedial measures for the volatile capital markets. This group had recommended the convening of an international conference on Money and Finance, on the same lines as the Bretton Woods Conference of the 1940's, to suggest a framework for a new monetary and financial system. The recent Monterry Consensus, produced by the expert group, set-up by the United Nations, had also urged the world leaders to evolve a new financial architecture for bringing about some semblance of discipline and predictability in the volatile capital flows and exchange rates. Feasibility of imposing a tax on the volatile capital flows, known as the Tobin Tax, has been debated since long. Unfortunately, no follow up has been taken up on the basis of the wisdom generated by these different expert groups. The result has been the recurrence of series of crisis situations and the consequent non-sustainability of the world financial and economic order.

Contours for a New Paradigm for Sustainable Use of Water

In order to bring home the point, that the problem of sustainability is a much wider issue, than tthe mere environmental sustainability, I have chosen two specific issues: The issue of sustainable use of water and the issue of Education and Governance for sustainable development. In this section, let me briefly introduce the contours of the challenges of sustainable use of water resources.

Projections of the water resources and the net use pattern emerging in the world indicate that water crisis is round the corner. Many urgent initiatives are required to deal with this severe crisis, which may throw the entire development process out of gear, if the water crisis is not managed in right time. The studies on this subject have identified the following steps, among others, to deal with this impending crisis:

- Water harvesting at the micro-level needs be intensified;

- Rain water harvesting and recharge of ground water and use of new technologies and practices, in this context, are urgently required;
- There is need for Integrated Land-Water management at the river basin level;
- Initiatives need to be taken for ecological-economic valuation of water at various levels;
- Many innovative steps need to be launched for the development of Technology and monitoring system for efficient use of irrigation water;
- Institutional framework for comprehensive assessment of large water projects needs to be strengthened;
- A new paradigm for sustainable use of water is feasible only when suitable principles for the pricing of water are evolved. There is need for more analytical and policy-oriented work in this field; and
- There is, as of now, hardly any Performance auditing of Water management programs. There is an urgent need for setting up of suitable capabilities and institutional framework for this purpose.

The above brief discussion brings out the point that our concern for sustainable development should go beyond the issues of environmental sustainability and deal with the challenges of many other dimensions of sustainability, in particular, that of sustainability of water resources.

UN Decade of Education for Sustainable Development

The subject of sustainable development is relatively new in the field of development economics. Both the theory and practice for sustainable development need to be still developed. It is for this purpose that some new initiatives are called for with the aim encouraging research, training and strengthening awareness about the challenges of sustainable development. It is in this background that the UN has declared the decade of 2001-11, as the UN Decade for Education on Sustainable Development. Unfortunately, very few are aware of this initiative of the UN system. It is for this reason that I thought of mentioning about this in this forum of researchers and practitioners. This UN decade has envisaged the following steps, among others, to meet its objectives:

- Education should include, Training, Teaching, Research, Grass-root level advocacy programs, etc.;
- Research requires multidisciplinary teams, consisting of Technology experts, engineers, scientists, economists, sociologists, political scientists, activists, and even philosophers. Such research should be launched on a massive scale;
- Governance for sustainable development is an essential prerequisite. For this purpose, a new cadre of Governance experts should be created along with suitable institutional framework for training them; and
- All universities should set-up Departments on Sustainable Development, and/or, at least, include this subject in the graduate and post-graduate and research curricula.

In this background, I would like to compliment Rabindra Bharati University for setting up this New Center on sustainable development and for conducting good innovative programs. I would like to assert that many more such initiatives are required throughout the country.

Ancient Indian Thoughts on Development—Some Insights

As mentioned in the beginning, I would like to dwell upon the unusual theme of ancient Indian thoughts on development, globalization, sustainability and related topics. My purpose is to bring out the point that such issues have occupied the minds of the thinkers in India from the very ancient times. My purpose is also to argue that many of the thoughts presented in our ancient literature have lot of relevance to the modern times and hence they need to be researched upon and put into practice, even in the present times.

Let me give, rather briefly, the characteristics of the Ancient Indian paradigm of development, as follows:

- Holistic Approach to Conceptual Framework and Analysis;

- Stress on a Blend of Materialism and Spiritualism; Stress on Human Factor;
- Stress on Duties as against Rights since the latter would be fulfilled when every one performs his/her duties;
- "Values" as the Prime Determinant of Human Welfare/ Happiness; Focus on *Values*, such as, contentment, caring, cooperation, trust, autonomy in thoughts, consistency, competence and commitment;
- Restraint on Self Consumption as the Prime Engine for Societal/Global Welfare;
- Social Ownership of Means of Production; Societal welfare should precede individual self-interest;
- Lifestyle of sustainable consumption as a modality for realizing Resource Balance in the Society;
- "Wisdom" is superior to Information and Knowledge; Absence of it, despite the possession of the latter, could be dangerous to the survival of mankind and global peace; and
- "Managing Oneself" as the prime starting point for evolving a Corruption Free, Efficient and Harmonious Social Order.

I would not like to elaborate on each of these points further. I would only like to mention that the principles laid down in this ancient Indian paradigm of development have great relevance for the modern times, to evolve a fair, equitable and welfare-oriented paradigm, for the maximum welfare of the mankind as a whole.

CONCLUDING REMARKS

Let me, at the end, present a synoptic overview of the issues and perceptions that have emerged as a result of our discussions.

- The World Order is now at Cross Roads; We are at a *turning point;* Search for a new paradigm of world order is urgently called for;
- The Process of Globalization as being practiced now is reaching the brink of collapse, like the previous Waves of Globalization; A new approach to globalization with equity and fairness, needs to be identified and put into practice;

- Sustainability is much wider than mere environmental sustainability. The debate on sustainability should cover all the different dimensions and evolve suitable strategies for sustainability in this comprehensive framework;
- A new Financial architecture is required to be urgently put in place;
- Concerns for sustainable use of water resources deserve special attention;
- Strengthening the institutional framework for education on the challenges of sustainable development is urgently required;
- Militancy, Terrorism, Collapse of Human Values, many regional wars on the Globe are symptoms of unsustainability of the present world order;
- A New Paradigm of Development, with a proactive state and focus on social aspects of development, physical and social infrastructure as also on the cultural environment and values, is required for realizing a sustainable process of development, in a Holistic Framework;
- Our Ancient Indian Thoughts on Development and Globalization provide many innovative insights that are relevant to the contemporary challenges of development. Efforts should be made to research on them, contextualize them for the challenges of the present times and integrate them with the current policy packages.

References

Bajaj, J.K and Srinivas, M.D. (Collated and Edited) (2008). Science Sustainability and Indian National Resurgence (A Collection of Speeches Delivered by Professor Murli Manhoar Joshi). Centre for Policy Studies, Chennai.

Deepak Nayyar (ed.) (2002). Governing Globalization, Issues and Institutions. (The United Nations University/WIDER), Oxford University Press, New Delhi.

Douglass C. North (2006). Understanding the Process of Economic Change. Academic Foundation, New Delhi (Published Under Arrangement with Princeton University Press, Princeton (USA) and Oxford (UK).

International Monetary Fund, Washington, D.C, (1974) Reprinted 1978. International Monetary Reform: Documents of the Committee of Twenty.

Jan Nederveen Pieterse (ed.) (2001). Global Futures: Shaping Globalization. Zed Books Ltd, New York, USA.

Panchamukhi,V.R. (1987). Planning, Development And The World Economic Order. Sangam Books Limited, U.K. By arrangement with Radiant Publishers, New Delhi.

Panchamukhi, V.R. and Rehman Sobhan (ed.) (1995). Towards An Asian Economic Area. Macmillan India Limited, New Delhi.

Panchamukhi, V.R. (2003). Fifty Years of Development Struggle: Some Reflections, ICSSR Occasional Monographs Series, No. 2, 2003.

Panchamukhi, V.R. (2000). *Indian Classical Thoughts on Economic Development and Management,* Bookwell Publication for Vidyaratna Sri R.S. Panchamukhi Indological Research Center, New Delhi.

Robbie Robertson (2003). The Three Waves of Globalization: A History of a Developing Global Consciousness. First Published by Zed Books Ltd, New York, USA. In Canda by Fernwood Publishing Ltd in 2003.

Shankar Acharya (2009). India and Global Crisis. Academic Foundation, New Delhi.

United Nations, New York and Geneva (2003). Management of Capital Flows: Comparative experiences and implications for Africa.

World Commission on Environment and Development (WCED), known by the name of its Chair *Gro Harlem Brundtland,* convened by the *United Nations* in 1983; Report of the Brundtland Commission, known as *Our Common Future.* Oxford University Press (1987).

Sustainable Development and Environment: Lessons from Ancient Indian Wisdom

S.D. Chamola

The present approach to economic growth and environmental protection is resulting in population growth, rising temperature, falling water tables, deforestation, pollution and the loss of many plant and animal species. The problem is so unsurmountable that our scientific and technical know-how on which we have relied so much is grossly inadequate to avoid the impending catastrophe which mankind has never faced in history.

Therefore, there is need of a paradigm shift and a fundamental change in our approach to sustainable development and environmental protection. In finding out an alternative model, many lessons can be learnt from the ancient Indian wisdom which can be incorporated in the suggested alternative model.

The Vedas exhort that the resources of the universe belong to all human and non-human beings. They have to be used with restraint and not for indulgence. There is a need to bring harmony between man and nature. The earth is our mother. It is our holy duty to protect her from degradation. Through the story of king Prithu and the cow the Bhagwat Purana concludes that for their better utilization, the resources of the earth should be under the trust of noble people who are devoted to the welfare of mankind and not in the hands of the rascals. The epics the Ramayana and the Mahabharata romanticize the forests, the rivers, the mountains and all beautiful things of nature and their utilization to enhance happiness of all.

The Buddhist and Jain chronicles teach that 'Ahinsa' or non-violance holds the key of harmony between man and other plant and animal species. Kautilya's Arthshastra holds that the earth is the key of all progress. Therefore, its acquisition and protection is the science of economics. Similar is the view of the Manusmriti. Both the authors have suggested a land-use planning which benefits both men and animals.

Even in recent centuries, there are many communities like the 'Bisnoi' community in Haryana and Rajasthan which has made tremendous sacrifices to protect trees and animals in their area as it is the basic tenet of their faith. Even in modern times Mahatma Gandhi has given an alternative model based on human values and materialistic growth. The basic formula is that there is enough on this earth for our need and shortage for our greed. Gandhian followers and environmentalists hold the view that inequalities, exploitation, terrorism, communal violence and many other ills in our society owe their origin to the present model of economic development of high growth rate at the cost of environmental degradation.

Present Unsustainable Development Strategy

The present model of economic growth is not harmonious with nature and consequently, economic development is resulting in degradation of environment. This strategy of economic development is not sustainable because for immediate gains we are becoming permanent losers. There are other reasons why the present mode of industrialization is not sustainable for India. First, the present model is violating people's fundamental rights of democracy and property and especially of the poors. Poor farmers are being ejected from their habitats without adequate compensation and rehabilitation. Their resources of land, trees, waterworks, and other common properties are being destroyed project after project. They are being deprived of their livelihood and socio-cultural milieu without making any alternative arrangement. Second, due to the present mode of industrialization, more people are becoming unemployed because the creation of jobs in the organized sector is very slow. There is no significant reduction in population dependent on agriculture and other rural households. The priority of this strategy is to keep the organized sector in fine trim even though the overwhelming majority of people in unorganized sector are neglected. This corporatisation-led growth strategy of creating wealth and its distribution, is based on highly faulty mechanism. The marketed-oriented forces act and react in such a way that the farmers and other sections of poor population are neglected and bypassed. This is resulting in steep inequalities in income and wealth. "There is no point in pretending to be an emerging super power with nearly half of our population in extreme poverty without minimum health care, sanitation, nutrition and education with largest number of illiterates and undernourished children."[1]

Ethical issues are also involved in the destruction and degradation of environment.[2] In the process of economic development, it is the moral duty of human beings to protect the rights of other non-human parts of the ecological system. Pain is an evil whether is inflicted on human or non-humans. All natural specie of fauna and flora including a lake, a wild river, a mountain, and even the entire biotech community has a right to have its integrity, stability and beauty preserved. It is also argued that possession of livable and clean environment is our birth right. Interfering in this right should be made punishable under the legal system. This should override people's property if they infringe upon our right to live in a clean and healthy way. Many thinkers have argued that the environmental crises are rooted in the social systems of hierarchy and domination that characterize our society. The social practices of racism, sexism, and social classes, economic exploitation go hand in hand with environmental destruction. Success is revealed in violation of rues and domination over other.

This domination is also exhibited in the destruction of environment for personal gains at the cost of society.

The present economic growth pattern also violates the rights of future generation,[3] for immediate gains denuding the earth of its valuable resources for future growth. Even there are sharp disparities in the utilization of these scarce resources among the various nations of the world.[4] For example, 6 per cent of the world population that lives within United States consumes 35 per cent of world's annual energy supplies whereas the 50 per cent of the world's people who inhabit less developed nations must get along with about 8 per cent in energy supplies. Each person in the United States in fact, consumes 15 times more energy than native South America, 24 times more than a native Asian and 31 times more than a native of Africa. American energy supplies are subsidized by the poor nations of the world.

Ancient Indian Thinking About Man and Environment

Indian civilization is one of the most ancient living civilizations. It has rich culture and literature unparalleled in any other such ancient civilization. The Indian concept of harmonious relationship between man and his environment has been explained in detail in its vast literature. For our present purpose, we shall be taking examples from its representative literary works. These include the Vedas, the Upnishads, the epics like the Ramayana and the Mahabharata, Buddhists and Jain chronicles, the Kautilya Arthshastra and the Manusmriti. This literature which is very relevant and widely read and practised over centuries together has valuable advice for the present civilization as how to harmonise relationship between economic development and environment on a sustainable basis.

Surprisingly, the basic under current running in all these literary works has remarkable similarity. Accordingly sustainable development implies a balance between human values and environmental protection. A balance has to be struck between spiritual and material thinkings, between greed and compassion, between self-restraint and self-indulgence, between violence and peace. In the Gandhian parlance, "on this earth there is enough for every one's need but not for their greed." Economic development is not an activity separate from other activities. The problem of man and nation can be resolved only by adopting a holistic approach rather than a fragmented approach such as economic and non-economic activities. The earth and its resources have to be worshipped and conserved and promoted. All plant and animal species are equal partners in sharing the resources of this earth. This is clearly proved by citing a few instances from the following literary works of ancient India.

THE VEDAS

Supporting this approach, the Atharva Veda, in its Prithvi Sukta, the Rishi prays that "on whom rest the ocean, the rivers, the water of wells, the tanks and the lakes on whom grow grains and other agricultural produce on whom exist all that breathe and move—let the earth place us also in the hands of the Lord who has already granted protection to those who deserve it, even before they are born.[5] Regarding the earth as the mother, the Rishi prays, "oh earth thou are my mother, I am thy son."[6]

The Ishavashya Upnishad goes further and says that "By one supreme ruler is this universe pervaded, whatever there is in this moving world is His. Hence consume the resources with restraint and renunciation. Don't covet and pounce upon these resources as vultures fall on dead body. After all, who is the owner of this earth?[7] The message is loud and clear that resources belong to all human and non-human beings. It is unethical to recklessly consume these resources for the greed of a few persons.

THE EPICS

The epics the Ramayana and the Mahabharata have romanticized the forests because their heros, i.e. Lord Ram and the Pandavas, spent a large and valuable part of their life in the forests destroying the evils and benefitted from the advice of the Rishis and Munis meditating there for the benefit of humanity. The Shanti Parvan of the Mahabharata gives valuable advice about development and the environment. In a dialogue Bheeshm Pitamah is advising the new king Udhishthir as how to run the administration. He says that "the earth is like a cow. It has to be ensured that it is not over milked. If the milk is shared properly between the calf and the milkman the calf becomes healthy and strong to carry the load. On the contrary an over-milked cow's calf becomes weak and cannot work efficiently. In the same way, by over exploitation of the earth by the King, the kingdom becomes poor and cannot perform well."[8]

The same point is proved by another analogy which says that, 'just as a honeybee sucks flowers and trees and does not destroy them', just a skilled dairyman milks the cow in such a manner that the calf is not affected and its odder is not crushed, in the same way the King should milk the earth of his empire."[9]

THE PURANAS

The Puranas which constitute an important part of our ancient Indian literature have cited many examples which show that environmental destruction occurs due to greed and selfishness of the people and the inability of the rulers who are poor and weak administrators to check this destruction. There is a story in the Shrimad Bhagwat Purana which states that a severe famine and drought occurred during the reign of King Prithu who otherwise was a pious and just king. The Lord asked the king to milk the earth who had turned into a cow. In the dialogue between the King and the cow shaped earth when the King asked the reasons of famine and drought, the cow-shaped earth replied, "My dear King not only are grains and herbs being used by non-devotees but as far as I am concerned, I am not being properly maintained. Indeed I am being neglected by kings who are not punishing those rascals who have turned into thieves by using grains for sense gratification. Consequently, I have hidden all these seeds which are meant for the performance of sacrifice."[10]

The moral of the story is that if the resources of the earth are consumed for personal interest and there is no compassion for other living entities, such evil people should be severely punished. The story further says that the cow-shaped earth put the conditions of its milking only by the Rishis and other spiritual people who are devoted to God and public welfare only then the earth was milked and the famine and drought ended. The lesson we can learn from this story is that natural resources should be under the ownership and guidance of pious-hearted rulers and not in the hands of those selfish and greedy rulers who are interested in using the public property for their selfish gains. The story says that the King Prithu agreed to the conditions of the milking the earth and the earth again became green healthy and put forth grains, and fruits for the subject. The Purans are replete with such stories and lessons which can he learnt both by the rulers and the subject.

THE BHAGWAT GEETA

The Bhagwat Geeta also reiterates that if man helps in protecting nature than nature will also protect man in return.[11] "The Lord in the beginning of universe created man and exhorted that so long you harmonise with nature then nature will bestow upon you the desired fruits. When man and nature make efforts for one another's prosperity, in the process both get prospered. Those selfish people who consume the fruits of nature alone without offering it to others, such a person is a thief and he is only consuming sin and not fruits."[12] Another message of the Bhagwat Geeta in this regard which is relevant relates to "Nishkam Karm" or actions without desiring fruits.[13] It implies selflessness as opposed to greed and selfishness. Applying this principle it is clear that senseless destruction of natural environment will stop if these resources are used not for selfish motive.

BUDDHISM AND JAINISM

Buddhism and Jainism, two major Indian religions started as a protestant movement against animal sacrifices in the Vedic rituals. Both these religions made "Ahimsa" or non-violance as their main area of thrust. Both of them emphasized right thinking and action. Both agree that the world is full of sufferings and that thirst, desire, attachment, etc., which are causes of wordly existence are the main reasons of our sufferings. The salvation lies in Ahimsa. There is violance of man against man and of man against plants and other species. Unless we emphasize self-restraint and control in our life the problem of violence and ecological destruction cannot be solved. Violence is the root cause of all our ills while pursuing economic development at the cost of ecology. Therefore, 'Ahimsa or non-violence is the only solutions. We must respect the existence of not only of majority of human population but also of our plant and animal species.[14]

KAUTILYA'S ARTHSHASTRA

Kautilya's Arthshastra is an authentic book on the ancient Indian polity. This book is supposed to be written around 3rd century B.C during the reign of Chandra Gupta Maurya. It is an Upveda of the Atharva Veda. It is an exposition of the Prithvi Sukta of the Atharva Veda given in chapter 12. A definition of the Arthshastra is found in the concluding section of the book. It says that Artha is the sustenance or livelihood of men; in other words, it means the earth is inhabited by man. Arthshastra is the science which is the means of the acquisition and protection of the earth.[15]

Kautilya was the first economist who emphasized accumulation of wealth as it is foundation of Dharma also.[16] All wealth comes from earth, therefore the acquisition and protection of the earth is the subject matter of economics. It is the duty both of the king and the subject to acquire and protect the earth. The state is expected to engage in various kinds of activities such as settlement of virgin land, building of dams, tanks and other irrigational work, providing postures for cattle, opening trade roots and ensuring safety on them working of miner and so on. He admitted that encroachment on pastures and agricultural land should not be tolerated. Provision has been made for the protection of forests, animal parks and specially elephant forests. Details of town planning are given by way of royal highways, construction of residential houses, healthy and clean environment, etc. Disaster management has been an important part of the Arthshastra. Fire, floods, famines, disease, rodents and wild animals are the natural disasters. Separate departments have been suggested to meet these disasters.[17]

MANUSMRITI

The Manusmriti in its present form is said to be composed during 200 B.C. to 200 A.D. The Manusmriti has given main objective of the book that the best means of securing welfare is to increase understanding procure fame and longevity which leads to extreme bliss.[18] According to Manusmriti, earth, water, air, fire and sky are the five elements of which the entire material world is composed. The heavenly bodies, the earth, birds, animal species, plants, human beings and all that exists in the ground, above the ground and under the ground, work in a harmonious manner. They maintain ecological balance. Man's encroachment on nature disturbs this equilibrium causing environmental hazards. These hazards result in floods, earthquakes, famines, epidemics, etc. Therefore, the only way of development is to maintain this harmony among the various of elements of life and equilibrium among both human and non-human beings.

According to Manusmriti a peaceful and harmonious development is possible only when Varna-Ashram Dharma is observed. There are four Varnas or social division consisting of Brahmin, Kshatriya, Vaishya and Shudra. Similarly, there are four Ashrams, i.e. Brahmcharya (celibacy), Grihsth (house-hold) Vanprastha (Departing to forests) and Sanyas (Renunciation of the world). The strict adherence to Varna Ashram Dharma holds the key of material and spiritual development and brings about harmony. Truth, non-violence, worship of God and nature, are the prescribed activities to attain economic prosperity and harmony with nature. If "Dharma is protected it will protect us and if Dharma is destroyed it will destroy us.[19]

The Bishnois

The Bishnoi community settled in Haryana and Rajasthan is famous for its article of faith in tree and wildlife protection. Unless this tree and wildlife is made an article of faith and an integral part of our religious rituals there is little hope for environmental protection. How this can be practiced should be learnt from the Bishnoi community. A Bishnoi knows that by protecting the green trees he is not doing any favour to any body. He is only protecting his life and securing the future of his own children.

At Khajarali, a village not far off from Jodhpur in Rajasthan, the memorial raised in the memory of the people who voluntarily secrified their lives for the protection of the Khejadi trees lists as many as 363 martyrs including men, women and children both young and old. This is perhaps unparalleled in the history of mankind. There is saying that if you happen to see a blackbuck, Chinkara or a common deer frolicking somewhere in the fields or in the village wasteland, no guess is needed you are surely in Bishnoi village. Their love for trees, blackbuck and Chinkara are almost a region, some shocking, some unbelievable for those who are alien to their faith and culture.[20]

MAHATMA GANDHI

The life and work of Gandhi have had a considerable influence on the contemporary environmental movement in India. From the Chipko Andolan to the Narmada Bachao Andolan environmental activists have relied heavily on Gandhian techniques of non-violent protest and have drawn abundantly on Gandhi's polemic against heavy industrialization. Again some of the movements' better known figures, for example, Chandi Prasad Bhatt, Sunderlal Bahuguna, Baba Amte and Medha Patekar have repeatedly underlined their own debt to Gandhi.

Gandhi anticipated our environmental concerns. His philosophy gives us an alternative perspective on development. While explaining how the current mode of development is exploitative of man by man and of nature by man, Gandhi's reservations about the wholesale industrialization of India are usually ascribed to moral grounds namely the selfishness and competitiveness of modern society but they also had markedly ecological undertones. The natural resources of country-side have been increasingly canalized to meet the needs of the urban industrial sector. The diversion of the forests, water, etc. to the elite, having accelerated process of environmental degradation has deprived rural and tribal communities of their traditional rights of access and use.

For Gandhi the distinguishing characteristic of modern civilization is an indefinite multiplicity of wants whereas ancient civilization were marked by imperative restriction upon and a strict regulating of these wants. In uncharacteristically intemperate tones, he spoke of his whole-heartedly detesting this desire to destroy distance and time to increase animal appetites and go to the ends of the earth in search of his satisfaction. If modern civilization stands for all this I call it satanic (*Young India*, 17.03.1927); one of his best known aphorism that the "world has enough for everybody's need but not enough for everybody's greed.[21]

Some Lessons

1. The present strategy of economic development is unsustainable because it is anti-poor and exploitative in nature. It violates of rights of other tree and animal species as well as of future generation. It is creating high disparities.
2. According to the ancient Indian thinking a balance has to be struck between spiritual and material thinking, between greed and compassion, between self-restraint and self-indulgence, and between violence and peace.
3. There is need to adopt a holistic approach rather than fragmented approach of economic and non-economic activities, separately.
4. The Vedas preach that the resources of the earth belong to all species and man should consume these resources with restraint and not recklessly for short-term gains.
5. The advice of the epics is to use the resources in such a way that all are benefited and no one is deprived.
6. According to the Bhagwat Purana the policy of using earth's resources should be vested with pious and selfless people and the King should be strong enough to punish the rascals who overexploit resource for their private motive.
7. The message of the Bhagwat Geeta is that if man promotes nature, in turn, nature will also promote man. "Nishkam Karmas" can solve our problem of destruction and degradation.
8. The Buddhist and Jain faiths are unanimous that Ahimsa or non-violence towards all other men and species is the only panacea of our present ills.
9. Kautilya's contention is that the acquisition and protection of the earth is the science of economics. A separate department should be established to deal with the problems.
10. The Manusmriti holds the view that any imbalance in the five elements of life creates disasters. Strict observance of Varna-Ashram Dharma is the only path of smooth development of man as well as of nature.
11. According to Bishnoi faith, unless trees and wildlife is

made an article of religious faith, environmental problem cannot be solved.

12. Mahatma Gandhi was of the firm view that "The world has enough for every body's need but not enough for everybody's greed.

NOTES AND REFERENCES

1. Bhaduri Amit and Medha Patkar (2009), Industrialisation for the People, of the People; *Economics and Political Weekly*, January.
2. Velasquez Manual G. (2009), Business Ethics: Concepts and Cases. Pearson Education Pvt. Ltd. Delhi.
3. Richard T. Degeorge (1973), The Environment Rights and Future Generation in K.M. Sayers eds. Ethics and Problems of the 21st Century.
4. Mishan, E.J. (1977). The Economics Growth Debate: An Assessment, George Allem & Unwin Ltd. , London.
5. यस्यां समुद्रः उत्सिन्धुः आपः यस्यां अन्नं, कृष्ट्यः सब मूदु!। यस्या इंद जिन्नन्ति प्राणांत् सानो भूमिः पूर्वपेये दधातुः

   Atharva Ved 12.3
6. माता भूमिः पुत्रः अहम् पृथिव्याः

   Atharva Ved 12.12
7. ईशावास्यम् इदं सर्वम् यत किचित जगत्यां जगत । तेन व्यक्तेन भुजीथाः मा गृधः कस्य स्विध नम्।

   Ishvashya Upnishad 1.1
8. भुतो वत्सो जातबलः पीडा सहति भारत, न कर्म कुरूते वत्सो मृशं दुग्धे युद्धिष्ठिर राष्ट्र मम्यति दुग्धं हि न कर्म कुरूतो मदेत।।

   Mahabharat 12.27.15
9. मधुदोह दुहेद राष्ट भ्रमर इव पादपम्। वत्सापेक्षी दुश्वैन स्तनांश्च न विकुटटयेत।

   Mahabharat 12.21.29
10. अपात्लिता नादृता च भवदिलोक पाल कैः चोरी भूतेऽथ लोकेऽह यज्ञाथगसमोषधी।।

    Shrimad Bhagwat 4.8.7
11. सहयक्षाः प्रजाः सृष्ठा पुरोवाच प्रजापतिः। अनेन प्रसंविष्यध्वमेष वैडस्विष्ठ कामधुक।।

    Bhagwat Geeta 3.10
12. यज्ञशिष्टाशिनः सन्तोमुच्यन्ते सर्वकिल्बिषेः मुन्जतेत त्वधं पापाये पचन्तयात्म काररात्

    Bhagwat Geeta 3.13
13. कर्मण्येवाधिकारस्ते मा फलेषु कदाचन। मा कर्मफलहेतुर्भूर्मा ते संगोऽस्त्वकर्मणि।।

    Bhagwat Geeta 2.47
14. Majumder, R.C. (1977), *Ancient India*; Motilal Banarasi Dass, Delhi.
15. मुनष्याशां वृतिरर्थः मुष्यवती भूमिरित्यर्थः तस्या पृथिव्या त्तामपालनोपायः शास्त्र मर्थशास्भमिति।

    Arthshastra 15.1.1-2
16. अर्थ एव प्रधान इति कौटिल्यः अर्थ मूलोहि धर्मकामाविति।

    Arthshastra 1.76-7

17. Chamola S.D. (2007), *Kautilya Arthshastra and the Science of Management*. Hope India Publication, Gurgaon.
18. Mansmriti 1.106.
19. धर्मएवं दृतोहन्ती धर्मोरक्षतिरक्षित, Manusmrit 8.25.
20. Chandla, M.S. (2000), *The Bishnois*, Aurva Publications, Chadigarh.
21. Ranga Rajau Mahesh (2007), *Environmental Issues in India, A Reader*. Darling Kindersley India Pvt. Ltd., Delhi.

Environment and Ecology in Kautilya's *Arthasastra*

Ratan Lal Basu

During the mid-twentieth century, rapid industrialization and indiscriminate use of non-renewable natural resources led to serious environmental pollution, ecological imbalance and rapid exhaustion of natural resources. Hazards like global warming, unpredictable climatic change, damage of ozone layer endangering the lives of not only plants and lower animals but the human beings also, alerted the scientists and environmentalists all over the globe.

As a result the "United Nations Conference on Human Environment" was held in 1972 at Stockholm (Sweden), and the matter gained further momentum after the "Earth Summit" held in 1992 at Rio de Janeiro (Brazil). Thus began the worldwide efforts towards preservation of environment and ecology and Acts in this regard were passed in most of the countries and efforts were made by both the governments and the NGOs to generate environment awareness among the masses.

However, long ago in ancient India, Vedic Samhitas,

Upanishadas, Epics, Dharmasastras, Arthasastras and various other texts prescribed measures for preservation of ecology and environment. Unlike the present day piecemeal and *ad hoc* approach towards the issue, ancient Indian environment consciousness was holistic in its approach and it sprang from the Upanishadic gospel *"Vasudhaiva kutumbakam"*, i.e., all the beings of the entire universe belong to the same family. Among the ancient Indian texts, Kautilya's Arthasastra is most secular and pragmatic in its approach as it was designed to specify rules which could be enforced by law by the king.

In this paper we are going to take up the major prescriptions in Kautilya's Arthasastra for preservation of environment and ecology and discuss their modern relevance.

I. Introduction

Recent awareness and in many cases fuss about environment and ecology arose only since the middle of the twentieth century. In fact, the current wave of environment awareness got a fillip only after the "United Nations Conference on Human Environment" held in 1972 at Stockholm (Sweden), and the matter gained further importance after the "Earth Summit" held in 1992 at Rio de Janeiro (Brazil). However during the early 20th century Rabindra Nath Tagore dealt in detail with various aspects of environment (and in a more coherent way than the present day piece meal analysis) in various writings.[1] In this regard he derived inspiration from ancient Indian texts. Vedic Samhitas, Upanishadas, Puranas, the Ramayana, the Mahabharata, Dharmasastras, Arthasastras, etc. embodied rules and regulations relevant to preservation of environment and ecology. Unlike the present day piecemeal and *ad hoc* approach towards the issue ancient Indian environment consciousness was holistic in its approach and it sprang from the Upanishadic gospel *"Vasudhaiva kutumbakam"*, i.e., all the beings of the entire universe belong to the same family. Among the ancient Indian texts, Kautilya's Arthasastra is most secular and pragmatic in its approach as it was designed to specify rules which could be enforced by law by the king. In this paper we are going to take up some important prescriptions embodied in the Arthasastra for preservation of environment and ecology. The rest of the paper is organized in the following manner:

Section-II: Prevention of Man-made Hazards; Section-III: Prevention of Natural Hazards and Section-IV: Relevance for Modern India

II. Prevention of Man-Made Hazards

Kautilya entrusted the task of protecting forests and other natural resources with the king [through different state officials]. He prescribed that appropriate plants should be grown to protect dry lands and pasture lands should be properly protected. The king should protect different types of forests, water reservoirs and mines.

To quote:

II/1/39: Thus the king should protect the product-forests, elephant-forests, irrigation works and mines that were made in ancient times and should start new ones.

II/2/4: And he should establish on its border or in conformity with the (suitability of the) land, another animal park where all animals are (welcomed) as guests (and given full protection).

II/2/5: And he should establish forests, one each for the products indicated as forest produce . . .

II/2/6: On the border (of the kingdom), he should establish a forest for elephants guarded by foresters.

II/2/7: The superintendent of the elephant-forest should, with the help of guards of the elephant-forest, protect the elephant-forest (whether) on the mountain, along a river, along lakes or in marshy tracts, with its boundaries, entrances and exits (fully) known.

II/2/8: They should kill anyone slaying an elephant.

Houses and other dwelling places, roads, cremation grounds, etc. should be properly constructed preserving environment. Every house should have proper arrangements for controlling fire. Any violation of the rules pertaining to these matters would make one liable to punishment. To quote:

II/4/1: Three royal highways running west to east and three running south to north, that should be the division of the residential area.

Every house should be constructed strictly on the basis of the rules [preserving environment] prescribed by the authorities. There should be proper arrangements in each house for sewage and proper disposal of wastes. The violators would be fined according to the gravity of the offence.

III/8/6: (He should make) the dung-hill, the water-course or the well, not in a place other than that suited to the house, except the water-ditch for a woman in confinement till the end of ten days (from delivery).

III/8/7: In case of transgression of that, the lowest fine for violence (shall be imposed).

III/8/9: He should cause to be made a deep-flowing water-course or one falling in a cascade, three *padas* away (from a neighbour's wall) or one *aratni* and a half (away).

III/8/10: In case of transgression of that, the fine is fifty-four *panas*.

III/8/11: He should cause to be made a place for carts and quadrupeds, a fire-place, a place for the large water-jar, the grinding mill or the pounding machine, one *pada* away or one *aratni* (from a neighbour's wall).

III/8/12: In case of transgression of that, the fine is twenty-four *panas*.

III/8/13: Between all two structures or two projecting rooms, (there is to be) an open lane one *kishku* (wide) or three *padas*.

III/8/14: Between them, the distance, between the eaves of roofs (is to be) four *angulas*, or one may over-lay the other.

III/8/15: He should cause to be made a side-door in the intervening lane, measuring one *kisku*, for making repairs to what is damaged, not (allowing) crowding.

I1I/8/16: For light, he should cause a small window to be made high up.

III/8/19: And he should cause that part above the verandah

which requires protection, to be covered by matting, or a wall touching (the roof), for fear of damage by rain.

In the above specifications Kautilya takes into consideration all aspects necessary for perfect harmony in dwelling places of the citizens.

Kautilya prescribes various fines and other punitive measures for polluting the environment by throwing dirt on the roads and highways or voiding urine and faeces at public places.

II/36/26: For throwing dirt on the road the fine shall be one-eighth (of a *pana),* for blocking it with muddy water, one quarter.

II/36/27: On the royal highway, (the fines shall be) double.

II/36/28: Fines for voiding faeces in a holy place, in a place for water, in a temple and in royal property are one *pana* rising successively by one *pana,* half these for passing urine.

But Kautilya was wise enough to realize that sometimes people arc compelled to commit the above mischief due to illness or similar reason. In that case the miscreant is exempted from fines. To quote:

II/36/29: If (these are) due to medicine, illness or fear, (the persons are) not to be fined.

Throwing carcasses or dead bodies at public places in the city are punishable by fines.

To quote:

II/36/30: For throwing the dead body of a cat, a dog, a ichneumon or a serpent inside the city, the fine shall be three *panas;* for throwing the dead body of a donkey, a camel, a mule, a horse or cattle, six *panas;* for a human corpse, fifty *panas.*

The dead bodies are to be cremated at cremation grounds only. Otherwise the offender will have to pay fines.

To quote:

II/36/33: For depositing burning (a corpse) elsewhere than in a cremation ground, the fine (shall be) twelve *panas.*

Kautilya also prescribes that every one should be careful about preserving common property and bio-diversity. Otherwise he would be fined.

To quote:

IV/10/4: In case of theft of deer or objects from deer-parks or produce-forests, (there shall be) a fine of one hundred.

IV/10/5: In case of theft of deer or birds (intended) for show or pleasure or in case of killing these, the fine shall be double.

No one should do anything to have harmful external effects on cultivation, irrigation system and other properties of other persons. Violators of this rule would be punished with fines. In case of setting fire to properties of others or common property, or bursting a dam containing water the punishment is death-sentence.

To quote:

III/9/27: In the case of damage to the ploughing or seeds in another's field by the use of a reservoir, channels or a field under water, they shall pay compensation in accordance with the damage.

III/9/28: In case of mutual damage to fields under water, parks and embankments, the fine (shall be) double the damage.

III/9/29: A tank on a lower level, constructed aflerwards, shall not flood with water a field watered by a tank on a higher level.

III/9/30: A (tank) constructed on a higher level shall not prevent the flooding with water of a lower tank, except when its use has ceased for three years.

III/9/31: For transgression of that, (the punishment shall be) the lowest fine for violence and the emptying of the tank.

III/10/5: For encroaching on a path for small animals or men the fine is twelve *panas;* on a path for large animals twenty-four punas, on a road for elephants or fields fifty-four *panas*, on a road to a dike or a forest one hundred and six, on a road to a cremation ground or a village two hundred, on a road in a *dronamukha* five hundred, on a road in a *sthaniya*, the countryside or pasture land, one thousand.

III/10/6: In case of reducing the size of these (roads), the fines are one-quarter of the fines (mentioned).

III/10/7: In case of ploughing on (them, the fines are) as prescribed.

IV/11/17: For one breaking a dam holding water, drowning in water at the same spot (shall be the punishment), the highest fine for violence if it was without water, the middle if it was in ruins and abandoned.

IV/11/20: He shall cause to be burnt in fire one who sets on fire a pasture, a field, a threshing ground, a house, a produce-forest or an elephant forest.

III. Prevention of Natural Hazards

Damage to environment may be caused also by natural hazards. All of them cannot be prevented by human material endeavours. But we may prevent some of them by our knowledge of science and coherent efforts. Kautilya's prescription for disaster management is worth-noting, particularly the anticipation of disasters and prior preparation for preventing them as far as possible. First he classifies the disasters caused by nature.

To quote:

IV/3/1: There are eight great calamities of a divine origin: fire, floods, disease, famine, rats, wild animals, serpents and evil spirits.

IV/3/2: From them he should protect the country.

Some of the important measures as prescribed by Kautilya to prevent natural hazards are discussed below.

Fire Hazards

The responsibility, of controlling hazards from fire and devising rules for the citizens so as to minimize hazards from fire, lies with the City Superintendent.

To quote:

IV/3/4: Prevention of fire is explained in 'Rules for the City Superintendent' and in connection with royal possessions in 'Rules for the Royal Residence'. Remedies against fire in residential areas and punitive measures against violators of fire-prevention rules are prescribed in the following *slokas*.

II/36/15: And (citizens shall take) steps against (an outbreak of) fire in summer.

II/36/16: In the two middle quarters of the day, one-eighth (of a *pana*) is the fine for (kindling) fire.

II/36/17: Or they should do their cooking outside (the house).

II/36/18: One quarter (of a *pana* is the fine) for not providing five jars, also a big jar, a trough, a ladder, an axe, a winnowing-basket, a hook, a 'hair-seizer' and a skin-bag.

II/36/23: For the owner, not running to save the house on fire, the fine (shall be) twelve *panas*, six *panas* for a tenant.

II/36/24: In case of (houses) catching fire through negligence, the fine (shall be) fifty-four *panas*.

The role of the City Superintendent as regards fire prevention is described in the following *slokas:*

II/36/19: The (City-superintendent) should remove things covered with grass or matting.

II/36/20: He shall make those who live by (the use of) fire reside in one locality.

II/36/22: Collections of water-jars should be placed in thousands on roads and at crossroads, gates and in royal precincts.

Flood Hazards

In the following *slokas,* Kautilya prescribes measures for prevention of hazards from flood-situations:

IV/3/6: In the rainy season, villages situated near water should live away from the level of the floods.

IV/3/7: And they should keep a collection of wooden planks, bamboos and boats.

Kautilya also emphasizes on mass-participation in rescue works. This is done both by moral suasion and legal measures.

To quote:

IV/3/8: They should rescue a (person) being carried away (by the flood) by means of gourds, skin-bags, canoes, tree-stems and rope-braids.

IV/3/9: For those who do not go to the rescue, the fine is twelve *panas*, except in the case of those without canoes.

Famine Hazards

The role of the state in famine management, as prescribed by Kautilya, is noteworthy. The ruler should have prior preparation for anticipated famines so that he is not caught napping. When the disaster actually occurs, he should take quick measures so as to minimize the harmful effects on the masses. The most interesting aspect in these prescriptions is that the ruler should relinquish if he fails to handle the famine situation.

To quote:

IV/3/17: During a famine, the king should make a store of seeds and food-stuffs and show favour (to the subjects), or (institute) the building of forts or water-works with the grant of food, or share (his) provisions (with them), or entrust the country (to another king).

Help of friendly foreign governments may also be sought if it is not possible to manage disaster by the efforts of the government of the affected country alone.

To quote:

IV/3/18: Or, he should seek shelter with allies, or cause a reduction or shilling (of the population).

In fact ecological and environmental awareness of Kautilya can be found in almost all the chapters. It is clear from the above discussion that Kautilya was much concerned about matters pertaining to the preservation of environment and ecology. To this end he prescribed various rules and also the punitive measures for violation of such rules. In this regard Kautilya's approach was holistic as he considered preservation of environment and ecology as integral parts of human living.

IV. Relevance for Modern India

As a follow-up of the Stockholm International Conference on Environment in 1972 several international agencies were formed for the protection of environment and ecology of Earth, Mentionworthy among them are:

1. Convention on International Trade in Endangered Species (CITES)
2. Environmental Protection Agency (EPA)
3. United Nations Environment Program (UNEP)
4. South Asia Co-operative Environment Program (SACEP)

This wave of environment awareness also encouraged the Government of India to pass various Acts to protect environment and ecology. Noted among these Acts are:

1. Water (Prevention and Control of Pollution) Act, 1974
2. Air (Prevention and Control of Pollution) Act, 1981
3. Environment (Protection) Act, 1986
4. Motor Vehicles Act, 1998

For implementation of the provisions of these Acts, "The Central Pollution Control Board" at the central level and "State Pollution Control Boards" at the state levels were formed in 1988. "The Department of Environment, Forest and Wild Life" was established in 1985. Mentionwothty among the various Government Agencies, and NGOs [Non-Government Organizations] concerning environment protection in India are:

1. Botanical Survey of India,
2. Zoological Survey of India,
3. National Wasteland Development Board,
4. Central Ganga Authority,
5. Bombay Natural History Society,

6. Central Forestry Commission,
7. Department of Non-conventional Energy Sources, and
8. Tata Energy Research Institute. *(Santra* (ed.) 2001)

One major shortcoming of all these measures [both at the national and at the international level] is that they are mainly *ad hoc* in nature and lack holistic approach. In fact, these measures are designed to solve specific problems and do not spring from any basic cosmological and holistic view. So, it has become difficult to coordinate and harmonize various types of activities and agencies designed to protect environment and ecology. These approaches may solve some specific problems pertaining to environment and ecology but would fail to remove the basic cause [which lies in our world outlook and view regarding nature and our relationship with it], which lead to the generation of damage to environment and ecology. In this regard we may learn from the ancient Indian texts like Kautilya's Arthasastra. We are to consider ourselves as a part of the harmonious universe and therefore regulate our material activities in accordance with the rules of nature. This approach towards environment and ecology is likely to generate deeper awareness about these matters among the common people so that they will be more willing to spontaneously co-operate in eco-preserving drives of the government or the NGOs. Kautilya's detailed specifications point out how every measure towards preservation of ecology and environment are coherently integrated.

It is true that environmental and ecological problems are now more severe and complicated than those at the time of Kautilya. Many aspects of environmental problems were totally absent in those days. But if we look deeply into the matter we will realize that all man-made environmental and ecological problems arise because of our greed or ignorance that make us commit things contrary to the balance of nature. The root cause of ecological problems is this and therefore same for all ages.

In Kautilya's age, moral preachings were not enough to prevent people from committing eco-damaging activities. So, he specifies punitive measures by the state for violating rules intended to maintain environment and ecology. All these approaches differ from the present *ad hoc* approach in the sense that we today do not go into the essence of the matter, and we treat every problem in isolation but not as the integral part of our living. Another important aspect of Kautilya's prescriptions is that he considers rescue of people from disasters a moral obligation of the ruler and he has no moral right to continue ruling if he fails to manage the disasters in the best possible manner. For example, in connection with famine control, Kautilya opines that in case it is not possible for the king to manage the disaster in an appropriate manner he must relinquish and entrust the country with another more competent king [see *Sloka* IV/3/17 above]. The modern rulers are to learn from this prescription of Kautilya.

Thus we see that the ancient texts may not help us in solving specific ecological and environmental problems, but they are going to give us deeper insight into the matter, which will help us solve our problems in a more harmonious way. This is the relevance of the ancient texts like the Arthasastra of Kautilya for modern India and, for that matter, the modern world.

Note and Reference

1. See Rabindranath Tagore: *Rabindra Rachanavali*, 125th Anniversary Edition:
   (i) Vol. 2: *Bharatvarsha*, p. 695;
   (ii) Vol. 6: *Swadeshi*, p. 497; *Samaj*, p. 517; *Bilaser Fans*, p. 526; *Shiksha*, p. 563.
   (iii) Vol. 7: *Dharma*, p. 447; *Shantiniketan*, p. 521; *Tapavan*, p. 690.
   (iv) Vol. 14: *Palli Prakriti*, p. 351; *Aranya Devata*, p. 372.

References

Santra, S.C. (ed.) (2001): *Current Perspective of Environmental Science*, Department of Environmental Science, University of Kalyani, Nadia, West Bengal.

Kangle, R.P. (1986): *Kautilyan Arthasastra*, Part-II (English translation), Motilal Banarasidass, Delhi. [All quotations from Arthasastra in our study are from this text. In quotations, 1/5/1 means Book-I, Chaptcr-5, *Sloka*-1 etc.]

ENVIRONMENT AND SUSTAINABLE DEVELOPMENT

M. ANISUR RAHMAN

Natural environment is the habitat of all living beings of the biosphere. Safe water, air and nutrients are necessary for their survival, growth and development. Every creature needs shelter to avert extremes of climatic condition and predators. Man is intellectually superior and he dominates over all the organs of the biosphere. He derives benefit from the environment and enjoys the natural resources and takes steps towards the development. Growth of human population stressed upon the exploration of natural resources and setting up of industries. Increased food production is the demand of population explosion. Extensive as well as intensive agriculture have been undertaken to feed the growing population. By doing so, the environment is going to be made uninhabitable by the production of wastes and detrimental substances around the human settlements and industries. These not only affect on the health and shorten the life span but also put questions on our survival.

Green house effect on the stratosphere makes a way to global warming affecting on our climatic condition to make our live unstable. The accumulations of green house gases are manifestations of developed nations for manufacturing modern amenities of life. They are to amend these or to bear responsibilities to clean up substances detrimental to life. Sustainable development is the demand for better and longer life on earth. The whole world should focus towards it. Otherwise, we have no way than to perish.

We need to maintain biological diversity for heterogenization. This is a key to survival of the race in any natural calamity or vagaries of weather. Loss of bio-diversity shall lead to desertification. Biological control rather than insecticidal or pesticidal control is to be allowed for a variety of life forms to flourish on economic basis. We may bend but not to break the ecosystem for continuity of life on earth.

INTRODUCTION

All living beings, including man live in reciprocation with their own environment. Environmental factors have great roles on growth, development, health and welfare of the living beings since time immemorial. Man has made efforts to modify his surroundings and dwelling habitats to alleviate adverse effects of extremes of temperature, sun, shade, excessive showers and predators. Agricultural and industrial developments are manifestations of manipulating nature and taping natural resources for better quality of life. Creation of new facilities, medical care and housing, safe drinking water and better food enhanced life expectancy. Infant mortality has been much reduced. Thereby human population increased to manifold. Although birth rate is in control in developed countries, it has been increased alarmingly in developing and under-developed countries. In order to sustain this large population of human beings that still continues to rise alarmingly, lots of land for housing, food, fuel, timber industries are necessary. Not only these, we are in need of more and more coal, petrol, gas, metal ores, cement, fertilizers, insecticides, pesticides, safe drinking water and what not. However, there are limitations of natural resources. Some resources are renewable but not all. We, therefore, cannot continue to expand our population and requirement of natural resource infinitely. It is now the time to draw a demarcation in our demand; otherwise we will perish in short time.

The quality of our environment has been constantly degrading and the stocks of non-renewable resources are decreasing fast. Exploitation of renewable natural resources at a faster rate, non-conservation of wild life, forest wealth, loss of gene pool, depletion of fossil fuel (coal, oil and gas), mineral wealth, etc. shall shorten human life on earth. On the other hand, pouring of industrial wastes and toxic chemicals, including non-biodegradable materials in the soil, water and air shall lead human habitation impossible. Destructive war weapons in the name of security and defense are degrading the environment and affecting badly the interest of the present and the future generations by defying the humanity. Therefore, we need to review the environmental issues whole heartedly for balanced development without serious destruction of our habitats and degradation of sum total environmental achievements. The twentieth century may be regarded as the period of maximum pollution, environmental degradation and fast diminution of biodiversity (*Ambasht & Ambasht*, 2006). Ecological balance must be restored and maintained, even if necessary, by promulgating international rules to safeguard biodiversity and by creating environmental awareness.

ENVIRONMENT

The earth we live in is of ours and of our forefathers. We therefore, have inherited it. The environment is an integral part of the earth and consisted of lithosphere, hydrosphere, atmosphere and the biosphere. In all the spheres the basic input of energy comes from the sun. The solar energy is trapped by the green plants. Hence, photosynthesis and other biotic processes are dependent on solar radiation. Plants are vital for the environment (*Rahman*, 1993). They are the primary producers in the ecosystem. Therefore, it is evident that without plants human life is not possible on this earth. We are in the tropics, I mean Bangladesh and Indo-Pak sub-continent. Thereby sunlight is more or less uniformly distributed throughout the year. This is coupled with adequate rainfall following a good coverage of vegetation except a few places in India suggesting a strong interaction with water, soil and nutrient movements in the environment (*Subramanian,* 2007).

Oceans are the vast reserves of water, next the ground water and dikes, beels, haores, etc. Besides these, there has been made some water reservoirs mainly in India for irrigation in order to enhance crop production for the growing population. Fresh water reservoirs and rivers provide excellent solvent and the flowing river water carried a number of dissolved substances, such as dissolved oxygen required for aquatic vegetation, essential nutrients (phosphorus, nitrogen and others) for non-aquatic vegetation along with effluents and pollutants. Oxygen and CO_2 level in the atmosphere is maintained by the photosynthetic organisms. Due to more use of coal and deforestation accumulated CO_2 levels in the atmosphere, we talk about greenhouse effect in our present day environment. Other green house gases are carbon monoxide (CO) and Chlorofluocarbons (CFCs). Although SO_2 is not a green house gas, it has adverse effect in the atmosphere by causing *acid rain* due to acidification of water molecules in the atmosphere. Besides this, oxides of nitrogen and aerosols (fine participate materials floating in the air) result in acid rain. This is damaging in agriculture as well as for the whole biosphere.

GREENHOUSE EFFECT

Concentration of green house gases: Carbon dioxide, methane, oxides of nitrogen, CFCs and ozone form a mantle around the globe and that more heat and infra-red radiations are trapped in it causing *global warming*. Recent climatic models suggest that global temperature shall rise by 2-6°C during the next century, if we assume that carbon dioxide concentration in troposphere increase to 600 ppm. This is only a speculation. If such thing happens, many parts of the world, including 1/3rd of southern part of Bangladesh will go under one meter of seawater due to Sea Level Rise (*Huq, et. al.* 1994). However, it is a fact that world climatic condition is changing rapidly and unnoticed. Human activity is mainly responsible for the accumulation of green house gases in the atmosphere. Energy production and its use, intensive agriculture and centralized industrialization, maintenance of huge livestock production (fattening industries for meat supply), CFCs, land-use modifications and industrial production are some of the aspects of human activities which are responsible for this accumulation. However, these activities are very much linked with growth, development and prosperity.

Power generation, refrigeration facilities, transportation activities, extensive deforestation for human habitation, industrial growth, extensive agriculture are necessities of loday. Restrictions on these shall low down living standards and amenities and a consequent reduction in growth, development and technological advancement. Developed nations are to be made responsible for all the environmental deteriorations caused so far and demanding their responsibilities to clean up. Though much environmental awareness has been created in the recent past, the key issues are lost in the conflict and confusion and little effective efforts have so far been taken to redress the situation.

Ozone Depletion and Global Warming

In the troposphere green house gases provide an effective thermal insulation while in the stratosphere many of these gases are responsible for causing ozone depletion. In spite of mutual opposite thermal gradient which prevents a free gaseous exchange between the troposphere and stratosphere, some transfer of gaseous material does take place between the two layers. Because of massive accumulation of greenhouse gases and a longer life span in the troposphere, these molecules rise at random and reach the stratosphere where they undergo dissolution yielding chlorine atoms, hydroxyl ions and nitric oxides which react with ozone degrading it to oxygen, resulting in a consequential decrease in holding capacity of heat by the ozone layer.

Nitric oxide is produced in the stratosphere by nitrous oxide derived from the troposphere below. Chlorine atoms are derived from the disintegration of CFCs. Plenty of hydroxyl ions are produced by dissociation of water molecules due to ultraviolet radiations in the stratosphere. By the last 2-3 decades the depletion of ozone is estimated to be about 8%. Consequently, an increase in the solar radiations containing harmful ultraviolet rays has taken place creating further heating effect by sunrays. (See *Farmen et al*, 1985; *Asthana, D.K.* and *Meera Ashthana*, 1998).

FOOD PRODUCTION IN INDIA

The rapid rise in total food grain production in India was made during 1950-70 by use of modern technology of intensive agriculture. Extension of irrigation facilities, fertilizers and pesticides use augmented further food production in recent years to acquire self-sufficiency in food grain production (*Asthana and Asthana*, 1998). This was possible by Green Revolution. However, intensive agriculture is highly expensive. This rapid strides in food grain production resulted in degradation of soils and their erosion. The cultivation of a large number of traditional varieties of crop plants has been replaced by HYV. Thereby, those traditional earlier varieties tolerant to the fluctuating climatic conditions have disappeared or in the line of extinction. Can people afford high cost of development? If the answer is yes, then there is another question, how far or how long?

SUSTAINABLE DEVELOPMENT

Better life in developed countries requires power and material inputs in converse releases wastes which in turn damages the environment resulting in quick exhaustion of natural resources. However, life in under-developed and also in the developing countries strives to live on sharing and clamouring for the basic requirements and ignorance very often damages the resource base of life support in the environment. Sustainable development lasts longer. However, nothing is ever lasting. There should be a fair sharing of natural resources among the living beings. Every one should get at least the basic amenities of everyday life regarding food, clothing, drinking water, shelter, education and medical facilities without damaging other life forms and the environment. We have to learn to live in harmony with nature. The resources of the nature should be properly managed, distributed and utilized meticulously. Then these are sufficient for all living beings of the biosphere. In future, however, we need to reduce the growth rate of human population (*Khoshoo*, 1990). Further, life support system requires (a) rationalized husbanding of all renewable resources, (b) conservation of all non-renewable resources and prolonging their life by recycling and reuse, and (c) to avoid wasteful use of natural resources.

Sustainable agriculture may be achieved by trusting on soil microbes for regeneration of soil fertility (*Rahman*, 2002) and soil conservation. Next, the reliance on bio-diversity is important for crop protection and higher productivity in order to attain sustainable food security (*Nazrul-Islam*, 2004). Crop rotation, mixed farming, cultivation of diversified varieties, we hope, shall safeguard against specific pathogens, insects and pests. These will improve nutrient status of the soil reducing the necessity of inorganic fertilizers.

CONCLUSION

In conclusion, I would like to say biological diversity is very important in our development. Long settled farmlands, rapid expansion of agricultural field and urban settlements are root causes of hindrance in bio-diversity. We have to stop the process of homogenization and introduce diversity in farming system and urban settlements.

Modern agriculture replaces the diversity of natural ecosystem with a single species of crop plants. Large expanses of land sown with a single strain or a variety is in itself a risky venture. It requires costly management in terms of fertilizers and insecticides. Being uniform in nature and in their genetic organization, the whole population of crop plants tends to behave in the same way in the face of unforeseen pathogens or *vagaries* of weather which could destroy the entire crop in a single stroke. Fields sown from horizon to horizon without interruption, without wind breaks or obstructions to the flow of water or air *enhance* soil erosion. Excessive use of chemical fertilizers depletes the soil of organic material. Insecticides kill harmful and useful insects alike. The microbial community of the soil, already reduced due to poor organic matter content, is adversely affected. Soil microbial life loses its diversity. Desertification follows. Of course, we can grow plants in sands as well by using necessary chemical fertilizers but the cost will be unaffordable. A certain degree of homogenization is the basis of all agriculture no doubt. Only a given type of plant is favoured, others are suppressed or weeded out for *enhanced* productivity. But we must not push uniformity to a dangerous level. Otherwise our agricultural system may be put to unsustainable. Modern agriculture depends heavily on irrigation, chemical fertilizers and the use of pesticides and insecticides without which the production is impossible.

A fundamental change is necessary for food supply on a secure footing. We should avoid temptation of high yields; rather to infuse diversity in our agricultural system and maintain tolerance to wild life around us. Indian Board for wild life conservation and protection has already taken steps towards growing consciousness among the people for wild life (*Trivedi* and *Sudarshan*, 1994). Alligators, Mongooses, Jackles, Foxes and other scavenger wild animals are disappearing from the nature. Frogs and toads are not seen very much in the fields and swamps, even in the rainy season. These all are to be nurtured in order to maintain biodiversity in the nature.

Desertification of our Borendra Region in Northern Bangladesh has already been started by deforestation and by excessive use of underground water for large-scale irrigation for food grain production. A compensatory aforestation is not more than 10% which cannot stop desertification. Natural system though modest in their yields are self-sustainable. Their nurturing is not unaffordable. Insects and pests may be controlled biologically or by mechanical means. A variety of life forms shall flourish with ease and harmonic way on an economic basis. We should try to make the man dominated systems as mimic natural systems, if humanity is to strive and survive longer in a healthy state.

References

Ambasht, R.S. and P.K. Ambasht, 2006: *Environment and Pollution*, Fourth Edition (Reprint), CBS Publishers and Destributors, New Delhi; Bangalore, India.

Asthana, D.K. and Meera Asthana, 1998: *Environment: Problems and Solutions*. S. Chand & Co. Ltd. New Delhi, India.

Huq, S., Ali, S.I. and Rahman, A.A., 1994. 'Sea Level Rise and Bangladesh.' *Environment and Development in Bangladesh*, Vol. I (ed). A. Atik Rahman, Ranna Haider, Saleemul Huq and Erik G. Jansen. The University Press Ltd. Dhaka.

Khoshoo, T.N., 1990, 'Man in Nature—Past, Present and Future', *Environmental Education and Sustainable Development*, Desh Bandhu, Harjit Singh, Maitra, A.K. (Ed), Indian Environmental Society, New Delhi.

Nasrin, S. and Rahman, M.A., 2007, Isolation and Characterization of Rhizosphere Bacteria and their effect on germination of rice seeds and growth of seedlings. *Journal of Pro-Science*, 15:77-82.

Nazurl-Islam, A.K.M., 2004, Biodiversity of Sustainable Food Security in Bangladesh. World Food Day, 18 October 2004, Bangladesh Agricultural Research Council, Dhaka.

Rahman, M.A., 1993, Plants for the environment. Presented at the 7th conference of the Botanical Society of Bangladesh, Dhaka.

Subramanian, V., 2007, *A Textbook in Environmental Science*, Third Reprint. Narosa Publishing House, Pvt. Ltd., New Delhi, Chennai, Mumbai, Calcutta, India.

Trivedi, P.R. and Sudarshan, K.N., 1994, *Environment and Natural Resources Conservation*, Commonwealth Publishers, New Delhi, India.

SECTION II

SUSTAINABLE MANAGEMENT OF NATURAL RESOURCES AND ENVIRONMENTAL GOVERNANCE

Sustainable Management of Renewable Resources with Reference to India

Kausik Gupta and Sanchita Som*

Sustainable Development is often interpreted as sustainable management of natural resource when we consider issues related to sustainable livelihood of the stakeholders associated with extraction of natural resources. Both the Human Development Reports (HDRs) and the Millennium Development Goals (MDGs) focus on poverty, hunger and environmental sustainability. The relationship between poverty and environmental sustainability is especially important in a resource dependent economy like India to study issues related to sustainable livelihood. In this paper we would like to focus on sustainable management of renewable resources and its impact on livelihood with special reference to India. For this purpose the paper takes into account of issues like

The authors acknowledge with thanks for the comments they received from the participants of the seminar. However, they are solely responsible for any error that may remain in the paper.

sustainable forest management and sustainable fishery management. Regarding sustainable management of forestry we find that India is experimenting with diverse programmes for protection, regeneration and biomass production in forests. The present paper attempts to examine how far these programmes are successful for sustainable management of Indian forests. Finally, the paper takes into account issues related to sustainable livelihood in connection with shrimp-mangrove system in the Sunderban region of India. It suggests some policies that can lead to sustainable development in the Indian shrimp-mangrove system.

Key Words: Millennium Development Goals, Community Forest Management, Shrimp-Mangrove System.

1. INTRODUCTION

Ecologists and environmental economists have interpreted the term sustainable development in different directions. The most common and the old definition of sustainable development has been given by The Report of the World Commission on Environment and Development (WCED) in 1987. It states that sustainable development is development that meets the need of the present without compromising the ability of the future generations to meet their own needs. Different environmental economists have considered various other definitions of sustainable development at different levels. Broadly speaking, sustainable development is one of the many processes of economic development. It deals with enhancement of living conditions of all generations. To achieve this use of resources will have to be appropriately designated over generations. In this paper we shall focus on sustainable management of renewable resources rather than confining ourselves to the definition of sustainable development itself. Sustainable management of natural resources focuses on sustainable livelihood of the stakeholders associated with extraction of natural resources.

In order to examine such sustainability from the point of view of sustainable resource management we next focus on the definition of sustainable development as shown by Ascher and Healy (1990). They have defined sustainable development as a process that serves the common interest through the attainment of four goals. Firstly, to achieve high per capita consumption that is sustainable for an indefinite period. It implies an optimal rate of use of natural resources over time. Secondly, there is existence of distributional equity in the society. Thirdly, there is prevalence of environmental protection including protection of biological diversity. Finally, there should be people's participation of all sectors of society starting from the grass-root level in decision-making. In this paper we would like to focus on sustainable management of renewable resources and its impact on livelihood with special reference to India. For this purpose the paper takes into account two important issues like sustainable forest management and sustainable fishery management.

The motivation behind this study originates from the fact that in India almost 28% of the rural and urban poor combined lives below the poverty line (*Economic Survey*, 2007-08) and proper sustainable management of renewable resources help the poor to uplift themselves above the poverty line. Apart from this as both the Human Development Reports (HDRs) and the Millennium Development Goals (MDGs) focus on poverty, hunger and environmental sustainability, a study on the relationship between poverty and environmental sustainability is especially important in a resource dependent economy like India to study issues related to sustainable livelihood. This can be captured when we focus on sustainable management of renewable resources in the context of India. For illustrative purposes we have selected sustainable management of two important renewable resources in Indian context: forestry and fishery.

The plan of the paper is as follows. Section II deals with a brief overview of the relationship between poverty and environmental sustainability. Section III deals with issues related to sustainable forestry management in the Indian context. Section IV deals with the issues related to sustainable fishery management in the Indian context. Finally, the concluding remarks are made in section V of the paper.

II. Poverty and Environmental Sustainability: A Brief Overview

Poverty and Environment are inter-linked through four main

dimensions: livelihood, resilience to environmental risks, health and economic development.[1] Poor people are often seen as compelled to exploit their surrounding for short-term survival or benefit and are assumed to be responsible for natural resource degradation. The relationship between poverty and environment has been captured traditionally by a very simple and circular framework and the relationship between these two issues is mediated by institutional, socio-economic and cultural factors. The concept of environmental entitlement is one of the major aspects in understanding the relationship between environment and poverty. Environmental entitlements are referred to two main attributes such as access to resources and control over the use of resources. Improving access to and control over environmental resources by the poor should provide a mechanism for the reduction of poverty. In the Indian context Jodha (1986) has shown that a large part of the consumption levels of rural poor are dependent on environmental resources. The great failures of the market and the state to provide appropriate modes of management of natural resources are responsible for environmental degradation. In this context National Environment Policy (NEP), 2006, has focused on some objectives which are as follows: (i) Conservation of critical environmental resources which are essential for enhancement of livelihood and human well-being along with economic growth, (ii) intra-generational equity and ensuring that the poorer sections of the society have secured access to these resources, (iii) inter-generational equity, (iv) integration of environmental concerns in economic and social development, (v) efficiency in environmental resource use, (vi) good governance regarding management and regulation of the use of environmental resources, and (vii) enhancement of resources for environmental conservation through mutually beneficial multi-stakeholder partnerships among local communities, public agencies, academic and research community, investors and multilateral/bilateral development partners. On the basis of the objectives of NEP, 2006, our study attempts to throw some light on sustainable management of renewable resources and its impact on livelihood with special reference to India. In the context we have focused on two major issues such as sustainable forest management and sustainable fishery management.

III. Issues Related to Sustainable Forest Management in the Indian Context

To achieve the path of sustainable development forestry has a multi-dimensional role. If an innovative strategy for the forestry sector with the help of multi-stakeholder partnerships can be formulated then it will help to achieve sustainable livelihood. The fishery problem can always be linked with the relation between shrimp aquaculture and mangroves. Fishery has links with biodiversity, livelihood of the fisherfolk and obviously with sustainability.

Conservation of forests is a prerequisite for sustainable development. Degraded forests account for 41% of forest cover. In India in the year 2006 it has been found that nearly 7 crore tribals and 20 crore non-tribals are dependent upon forest. It implies rapid forest degradation in recent years. To conserve forest resources and to provide alternative livelihood for the forest dependents stress has been given on the programmes like forest department (FD) managed system, Joint forest management (JFM), Industry promoted forestry, Self-initiated forest protection group (SIFPG), Community forest management (CFM), etc.

JFM is a participatory forest management system between village community and the State Forest Department which came into effect from June 1990. It aims to bring the forest department and the local people together in the management of degraded forests. In this management system the rules and regulations regarding protection of forest, the formation of the village protection committee and the sharing of the forest products are laid down by the State Forest Department. From the point of view of village community the two basic types of local operational institutions are the Forest Protection Committee (FPC) and Eco-development Committee (EDC). In India, since 1990 all the states have been trying to bring more and more forest area under this system. West Bengal has been a pioneer of JFM. It originated at the Arabari Forest Range (*Sal* Forests) in West Medinipur near Medinipur town (in the state of West Bengal) as Arabari socio-economic project in the year 1971-72. It expanded to other parts of the state in 1987 (especially Southern West Bengal) and later on JFM was modified in the years 1990 and 1991. In the northern part of the state of West Bengal the forests do not have the same regenerative capacity compared to the forests of Southern West Bengal. The success of JFM in West Bengal is moderate. JFM has succeeded in Southern West Bengal but failed in Northern West Bengal because JFM in West Bengal is not fulfilling its poverty alleviation potential. It implies in northern West Bengal the FPC and EDC members do not have the same kind of incentive to participate in the management process as we find in the case of JFM in Southern West Bengal. West Bengal Government's Forest Department has de-linked livelihood development from forest development. In this programme less attention has been paid to non-timber forest product which is the major part of poor people's livelihood. Poorer sections of the society are not considered separately in this programme resulting in stagnation and disillusionment of JFM. In the northern part of West Bengal participatory forest management is more of CFM type than of JFM type. It is based on the operations of FPC and EDC. CFM is also popular in other states of India (like Orissa). CFM believes that forest protection does not require any external financial support. The required investment can be generated internally. So the task of the researchers is to settle the issue that what form of participatory forest management should be the appropriate one for India.

IV. Issues Related to Sustainable Fishery Management in the Indian Context

The fisheries sector in India is an emerging sector of our national economy that promotes income and employment generation and also acts as a foreign exchange earner. Since early 1990's India has experienced very fast growth in shrimp cultivation mainly due to growing demand for shrimp and attractive international shrimp prices. Over the past ten years, India has become the fourth largest shrimp producer in the world (Fig. 1). There are two types of fisheries in India: (1) inland fisheries, and (2) marine fisheries. Marine fishing is the traditional occupation of the fishermen living along the coastal areas of India and it also provides one of the major means of livelihood for a large section of people inhabiting in the coastal region of West Bengal. Inland fisheries include country's great river system, the network of irrigation canals, lakes, beels and ponds, etc. So it can

Fig. 1: Top Ten Shrimp Producers (Aquaculture), 1975-2002

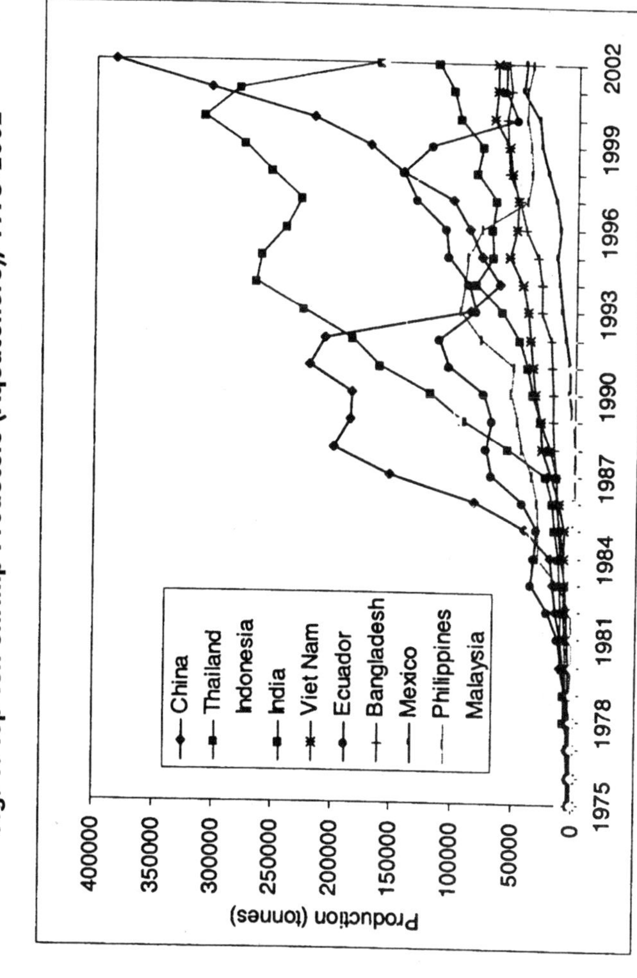

Source: FAO, Fishstat.

be conducted in a more controlled environment compared to marine fisheries and there are less seasonal variation in fish catches in case of inland fisheries compared to marine fisheries. To get a clear picture of the stakeholders associated with inland and marine fishery we consider a preliminary assessment of gainers and losers associated with both the above-mentioned two types of fisheries.

TABLE 1

Preliminary Assessment of Gainers and Losers in Inland and Marine Fisheries

Activity	*Gainers*	*Losers*
Measures to promote sweet water inland fisheries (conservation of sweet water inland wetlands)	The fishermen associated with fishery activities on ponds or tanks or lakes under the co-operatives, the landless labourers, the landowners, the full-time and part-time agricultural workers, the washer men who use ponds etc.	Fishermen engaged in open access river fisheries, the landowners and the full time agricultural workers.
Measures to promote brackish water inland fisheries owned by the Government (conservation of brackish water ponds).	Fishermen associated with brackish water ponds, the landless labourers etc.	The agricultural workers, the marine fishermen etc.
Measures to promote brackish water inland fisheries owned by the private sector units (conservation of brackish water ponds).	Fishermen associated with brackish water ponds, the landless labourers etc.	The agricultural workers, the landowners, the marine fishermen etc.
Measures to promote marine fisheries (through coastal zone regulations)	The marine fishermen, the boat owners, the government etc.	The aquaculture farm owners

From Table 1 it appears that separate analysis of Inland and marine fisheries is vast topic and it is beyond the scope of the present paper. In fact it is very difficult even to consider all the aspects of any particular type of fishery. So in the present paper we have confined ourselves to some of the important aspects of fishery management which cover some parts of marine fishery and some parts of inland fisheries. The ideal area for this punch is to focus on the inter-relationship among mangroves, capture shrimp production (part of marine fisheries) and aquaculture shrimp production (brackish water inland fisheries) in the Indian context with special reference to the Indian Sunderbans. This issue of fishery management is also important in the Indian context from the point of view of the linkages between biodiversity and sustainable livelihood.

Traditional fisheries can provide only a part of the total aquatic proteins demand, so rest of the part must come from aquaculture. Production of marine fisheries is less than production of inland fisheries but contribution to national income is less for inland fisheries. The total shrimp production of capture and aquaculture is shown in Fig. 2.

In the 10th five-year plan (2002-07), the Government of India has taken some steps to move the country towards sustainable aquaculture development. There are ten marine states in India namely Kerala, Andhra Pradesh, Maharashtra, Goa, Karnataka, Tamil Nadu, Pondicherry, West Bengal, Orissa and Gujarat. Among these States in Andhra Pradesh there exists hatchery-based cultivation and in the Sunderbans there exists local fry collection-based Cultivation.

Mangroves are a very important category of wetland systems that shelter coastlines and estuaries. They are rich in flora and fauna, especially in the tropics. Their major environmental services include storm protection, shore stabilization and control of erosion and flooding. They are also a biomass expansion and a nursery ground for marine life. One of the key aspects of mangrove ecosystem is the relationship involving shrimp and mangroves. On the ecological side, mangroves serve as the breeding and rearing area for matured shrimp stocks that are commercially important. They also harbour shrimp fry that are collected and supplied to the growing coastal shrimp aquaculture farms.

Along with the rapid expansion of the shrimp aquaculture

Fig. 2: Shrimp Production in India (Aquaculture and Capture), 1975-2002

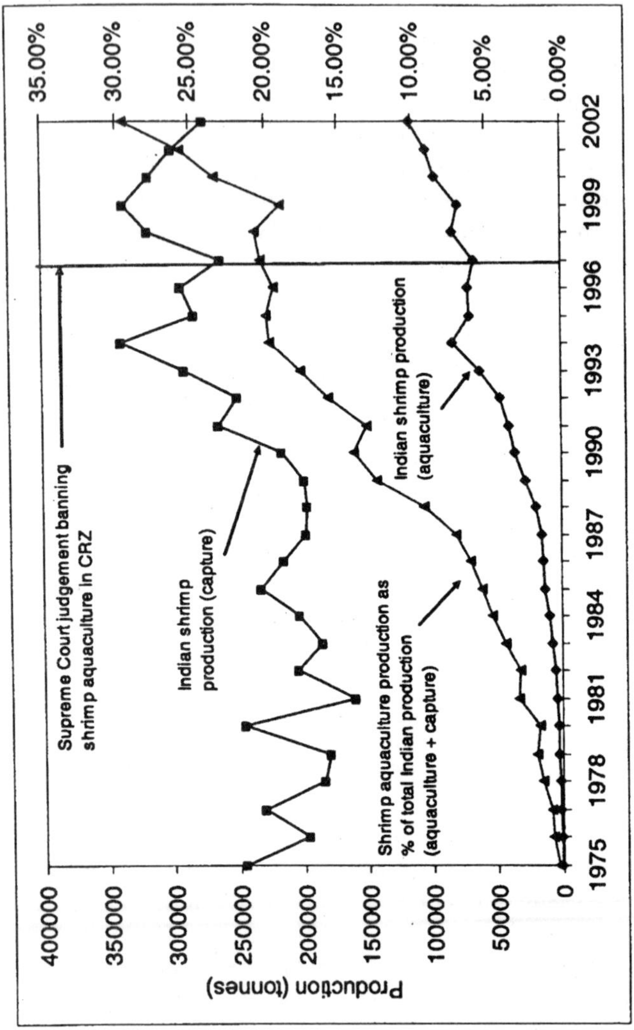

Source: FAO, Fishstat.

industry we find that in various countries a debate has emerged between proponents and detractors of this industry. In this regard we have to analyze the ecological and economic linkages between mangrove and shrimp. The detractors claim that expansion of aquaculture shrimp farm is responsible for degradation of mangrove forests (In case of Thailand, Indonesia). With the expansion of shrimp farm more and more exploitation of shrimp fry takes place and that creates a reduction in yield from ocean shrimp fishery. An increase in collection of shrimp fry is not highly remunerative for shrimp fry collectors. It also affects their health. So ultimately livelihoods of the fry collectors are affected. It hampers sustainability, which affects the maintenance of a steady standard of living. This is the common view regarding mangrove-shrimp relationship. But according to proponents this is not true for India, as data suggest. However, development of shrimp culture cannot be sustainable without mangrove. Various scholars found that aquaculture farming use already degraded and destroyed mangrove. But there exists a belief among many researchers that shrimp aquaculture has some negative externalities on environment but the main causes for mangrove destruction are human activities like urban development, timber extraction, etc.

Table 2 shows the distribution of mangroves in India on the basis of the report of the Ministry of Environmental and Forests

TABLE 2

Distribution of Mangroves in India

Sl.No.	States	Area in sq.km.
1.	Andaman and Nicobar Islands	1190
2.	West Bengal	4200
3.	Orissa	150
4.	Andhra Pradesh	200
5.	Tamil Nadu	150
6.	Karnataka	60
7.	Goa	200
8.	Gujarat	260
9.	Maharashtra	330
	Total	6740

Source: Ministry of Environment and Forests (MoEF), 1990.

in 1990. It shows that West Bengal has the highest area (in sq.km.) of mangrove coverage in India, followed by the Andaman and Nicobar Islands.

The common view regarding mangrove shrimp system has been shown in Fig. 3 in terms of a flow chart. It shows the common view how expansion of shrimp firms causes mangrove destruction and affects negatively capture fishery. The ocean fishermen are severely affected by expansion of shrimp farm as most of shrimp fry are collected to produce mature shrimp by the shrimp farm.

Fig. 3: The Shrimp-mangrove System

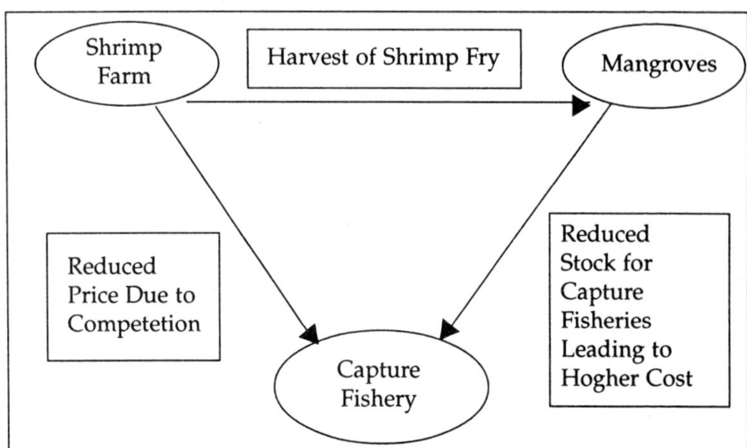

The literature on mangrove-shrimp linkages is quite extensive.[2] Most of the works focus on the benefits as well as the costs associated with conservation of mangroves. The researchers are interested in the ecological and economic linkages between mangroves and shrimp.[3] Again, some of the authors have focused on the institutional aspects of these linkages.[4] The other category of work deals explicitly with the problems of environmental valuation in the context of mangrove-fishery linkages.[5] Most of the authors have concluded that expansion of aquaculture shrimp farms is mainly responsible for degradation of mangrove forests in countries like Thailand, Philippines, and Indonesia.

The above-mentioned scholastic works on mangrove-shrimp linkages for the Southeast Asian countries has caused a belief among some of the researchers that in Indian coastal belt shrimp culture is the main destroyer of mangrove forests. Over the years this belief has been converted to a myth among various researchers and environmentalists who are interested in conservation of mangrove forests. In order to search for the reality regarding mangrove-shrimp linkages in India one should be careful in analyzing the data on coverage of mangrove forest area in the Indian coastal zones. The present paper attempts to search for the real picture on mangrove-shrimp linkages by analyzing the limited data that are available in this regard. Though shrimp culture is often regarded as the major destroyer of mangroves, its development is not as closely related to the availability of mangroves as often perceived. However, development of shrimp culture cannot be sustainable without mangroves.

On the basis of analysis of depletion of mangroves in India by various scholars we find that aquaculturists in India are more often than not the end users of already degraded or destroyed mangroves rather than the primary culprits of mangrove destruction.[6] There are different estimates of mangroves in India by various agencies like Ministry of Environment and Forests (MOEF). The National Institute of Oceanography (NIO), Goa, Forest Survey of India (FSI), Dehradun, etc. The paucity of information on mangroves in India has resulted in discrepancies between various estimates.

The time-series data show that maximum mangrove cover has been lost in West Bengal (31.15%). Following FAO (2002) we find that the traditional shrimp farms in West Bengal (*bheris* in local vocabulary) covering an area of approximately 33,000 ha were in operation much before 1987 and shrimp farms developed in West Bengal during 1987-94, covered an area of 15,444 ha only. On the contrary, the mangrove loss in West Bengal during the same period is reported as 2081 sq. km or 208,100 ha. Assuming the entire 15,444 ha of the new shrimp farms in West Bengal were developed in the mangrove area, the balance 192,656 ha loss of mangrove cover in West Bengal during 1987-94 must have been due to other human activities.

Studies conducted on Sunderban forests of West Bengal during the late eighties have reported that large areas of mangroves have gone for urban development, apart from being converted for agriculture, salt production and industrial uses.

In the year 2001 it was reported by the government of Andhra Pradesh that about 2838 hectares (ha) of mangroves have been converted for shrimp farming. However, if we examine carefully the data we find that a major chunk of mangrove in Godavari estuarine system was dereserved in the late 1970s when the importance of mangroves was not recognized fully.

The Indian coastline is about 5,700 km long. It can be divided into the east and west coasts and island chains. The east coast covers the maritime states like Tamil Nadu, Andhra Pradesh, Orissa, West Bengal and Andaman-Nicobar Islands. The West coast extends from Kerala, Karnataka, Goa, Maharashtra, Gujarat and also includes the coral atolls of Lakshadweep Islands. The total mangrove area along the Indian coast is estimated to be approximately 700,000 ha. (Among the total area 80% is along the east coast and 20% is along the west coast).

Considering 78,702 ha of area under shrimp farming in 2001 in the state, the 2838 ha of mangrove area accounts for only 3.6% of the total area under shrimp farming since 2001 [*FAO*, 2002].

On the basis of the study conducted by National Remote Sensing Agency (NRSA) in 1993 we find that in Guntur district of Andhra Pradesh the total mangrove loss during the 20 year period 1973 to 1992 was 952 ha of which 752 ha was lost during 1973-85. Only 200 ha was lost between 1985 to 1992 which is 5.0% of the total area of shrimp farming (4048 ha) during the period. In Krishna district, 405 ha of mangroves was lost during the said period and shrimp culture development was about 6005 ha [*FAO*, 2002].

On the basis of the work done by M.S. Swaminathan Research Foundation, Chennai, in the Pichavaram area of Tamil Nadu in 1996 we find that the degradation of mangroves is largely due to biophysical factors such as changes in topography and under water flow pattern. Further, remote sensing studies conducted by The Institute of Ocean Management, Anna University, Pichavaram, have shown that mangrove forests have increased from 461.98 ha in 1987 to 475.99 ha in 1998; while the degraded mangroves have reduced from 152.93 ha to 120.25 ha. During the same period, area under aquaculture increased from 6.99 ha to 331.27 ha.

As per recent information provided by some of the coastal state governments and union territory administrations, about 25.8 ha of mangroves have been converted into shrimp farms in Kerala, and no mangrove land has been converted into shrimp farms in West Bengal, Orissa, Gujarat, Pondicherry and Goa.

We can thus conclude that destruction of mangroves in India in recent years is mainly due to urban development, wood cutting, development of salt pans, agriculture and traditional brackish water farming and the destruction of mangroves due to commercial shrimp farms is insignificant. This is actually the reality.

5. Concluding Remarks

Various programmes have been taken for sustainable management of renewable resources. For example in 1986 the Environmental Protection Act has been implemented to provide a holistic framework for protection and improvement to the environment. The Wildlife (Protection) Act was introduced in 1972 and was amended in 1983, 1986, 1991 and 2002. The Coastal Regulation Acts and Biological Diversity Act, 2002 are also very important in this connection. However, it has been noticed that many problems are associated with the implementation of the existing relevant policies. The environmental challenges which were mentioned in NEP are more of general type and less of India-specific. Moreover, the policy proposal has stated for 'local recognition of traditional rights'. This is ambiguous unless the role of forest department in the context of its relationship with local communities is clearly specified. In India, for example, the state of West Bengal is not fulfilling its poverty alleviation potential through JFM. The forest department has delinked livelihood development from forest development in most parts of the state of West Bengal. The inadequate organizational structures acts as an impediment to procurement of non-timber forest products (NTFPs). Although NEP proposes to build up a 'community reserve' which encourage local communities associated with forestry to find alternative livelihoods but this is not properly implemented.

The main question that arises in the context of shrimp aquaculture in India is that whether it is sustainable or not. Philcox (2006) has argued that shrimp aquaculture is not sustainable in its current form in India when assessed against the four criteria stated by Ascher and Healy (1990). Philcox (2006) has argued that: (i) though substantial export revenues generate higher per capita consumption, it is by no means for an indefinite period nor does it imply an optimal use of natural resources over time. (ii) Shrimp farmers with access to capital, and middlemen face high costs associated with the negative social, environmental and economic externalities, which are borne by landless and marginalized communities. (iii) Participation by the grass-root people of the coastal communities in the decision-making process is minimal.

In order to pursue a sustainable development path the policy-maker should focus on some aspects. For example, the number of shrimp hatcheries should be increased to avoid the risk of white spot disease. Apart from this, initiative should be taken for generating alternative income opportunities for the fry collectors. By imposing taxes or by imposing licensing fee on the shrimp producers in this region funds can be raised for alternative income opportunities. If these funds are uilised properly for hatchery development then the threat of white spot disease will disappear. Initiative should be taken to develop cold storage system also. The environmental awareness needs to be increased among the Stakeholders or the various affected groups. Along with this Government should take the initiative to educate the shrimp farmers through proper training programmes, in applying shrimp fry and chemical inputs for matured shrimp production. Conservation of mangroves in the Indian part of Sunderbans can be promoted by following the joint mangrove forest management scheme.

The other policies for the well-being of the shrimp farmers of this region include formation of cooperatives or group aquaculture or aquaculture clubs as we find in Andhra Pradesh and Kerala. Formation of shrimp farmer's co-operatives or group aquaculture or aquaculture clubs will help to protect the interest of its members by checking exploitation from the shrimp merchants or the moneylenders. Even the members can be provided with insurance facility that is lacking in this region. All these policies will benefit the shrimp farmers not only in the short-run but will also help them in the long-run to achieve the sustainable level of livelihood.

If these policies are implemented efficiently, overexploitation of shrimp fry can be checked and it is expected that there will be a sustainable increase in production of matured shrimp in the near future.

NOTES AND REFERENCES

1. UNDP-UNEP Poverty Environment Initiative, Angela Lusigi, 2008.
2. Mangrove swamps are considered very suitable for shrimp farming because the areas are flooded with brackish stagnant water that is ideal for aquaculture
3. The most important work in this connection is the paper by Ruitenbeek (1994).
4. See for example the works by Nickerson (1999), Primavera (2000), etc.
5. For example the works by Gilbert and Janssen (1998), Naylor and Drew (1998), Janssen and Padilla (1999), Ronnback (1999), Barbier (2000), etc.

REFERENCES

Ascher, W., and R. Healy (1990): *Natural resource policy-making in Developing Countries*, Duke University Press, Durham.

Barbier, E.B. (2000): "Valuing the Environmental as Input: Review of Applications to Mangrove-Fishery Linkages", *Ecological Economics*, 35:47-61.

Duraiappah Anantha (1996): "Poverty and Environmental Degradation: a Literature Review and Analysis", *CREED, Working Paper Series* No. 8.

FAO (2002): *FAO Aqua Book-India*, Rome, Italy.

FAO (2000): *The State of World Fisheries and Aquaculture*, Rome, Italy.

Gilbert, A.J. and Janssen (1998): "Use of Environmental function to communicate the value of a Mangrove Ecosystem under different Management Regimes", *Ecological Economics*, 25:323-346.

Government of India (2002): *10th Five-Year Plan*. Planning Commission.

Gupta, K. (2005): "Mangrove-Fishery Linkages under Alternative Policy Regimes", Sengupta, N. and J. Bandyopadhyay (eds.), *Biodiversity and Quality of Life*, 272-93, Macmillan, New Delhi.

Gupta, K., J. Chowdhury and R. Khaddaria (2007): "Aquaculture *vs*. Wild Shrimp Fisheries: A Bio-Economic Analysis for West Bengal and Orissa". Pushpam Kumar and B. Sudhakar Reddy (ed.), *"Ecology and Human Well Being"*, 93-103, Sage Publications India Pvt. Ltd.

Gupta, K., J. Chowdhury and S. Som (2006): "The Aquaculture Shrimp Farm Model", Working Paper based on Node 1 of SHARP. *Working Paper No. GCP-JU-SHARP-3*, Global Change Programme, Jadavpur University, India. Available in http://www.sici.org

Janssen, R. and J.E. Padilla (1999): "Preservation and Conservation? Valuation and Evaluation of a Mangrove Forest in the Philippines", *Environment and Resource Economics*, 14: 279-331.

Jodha, N.S. (1986): "Common Property Resources and Rural Poor in Dry Regions in India", *Economic and Political Weekly:* 187-97.

Kowler, D., J. Roy, K. Gupta, W.K. De La Mare, W. Haider and A.C. Gupta (2006): *"Assessing Environmental Management Options to Achieve Sustainability in the Shrimp-Mangrove System in the Indian Coastal Zone of Bay of Bengal"*. Narrative Summary of SHARP.

Final Report, Shastri Indo-Canadian Institute, New Delhi (India) and Calgary (Canada). http://www.sici.org

Lusigi Angela (2008): "Linking Poverty to Environmental Sustainability", *UNDP-UNEP Poverty—Environment Initiative.*

Naylor, R. and M. Drew (1998): "Valuing Mangrove Resources in Kosrae, Micronesia", *Environment and Development Economics,* 3: 471-90.

Nickerson, D.J. (1999): "Trade-offs of Mangrove Area Development in the Philippines", *Ecological Economics,* 28: 279-98.

Philcox, N. (2006): "A Policy Review of Shrimp Aquaculture in India," Working Paper based on Nodes 1 and 3 of SHARP. *Working Paper No. 4,* School of Resource and Environmental Management, Simon Fraser University, *Canada.* http://www.sici.org

Primavera, J.H. (2000): " Development and Conservation of Philippine Mangroves: Institutional Issues", *Ecological Economics,* 35: 91-106.

Ronnback, P. (1999): "The Ecological Basis for Economic Value of Seafood Production Supported by Mangrove ecosystems", *Ecological Economics,* 29: 235-52.

Ruitenbeek, H.J. (1994): "Modelling Economy-Ecology Linkages in Mangroves: Economic Evidence for Promoting Conservation in Bintuni Bay, Indonesia", *Ecological Economics,* 10:233-47.

EPI AND ITS RELATIONSHIP WITH SOME MACROECONOMIC INDICATORS

AMRITA PAL AND JAYANTA SINHA

Over the past three decades, there is increasing consciousness about environmental problems facing communities, nations, and the world. During this period, natural resource and environmental issues have grown in scope and urgency. Sustainable development is a socio-ecological process characterized by the fulfilment of human needs keeping the quality of the natural environment indefinitely. In 1987, came the Brundtland Report with the most often-quoted definition of sustainable development, i.e. it is the development that 'meets the needs of the present generation without compromising the ability of future generations to meet their own needs'. It is very important to know the performance of a particular country with regard to the maintaining of a quality natural environment and to set target proposed in the UN Millennium Development Goals (UN MDG). The Environmental Performance Index (EPI) is a procedure of quantifying and numerically benchmarking the environmental performance of a particular country's policies.

The EPI was led by the Environmental Sustainability Index (ESI), published between 1999 and 2005. Up till January, 2008 two EPI reports have been prepared and released—the Pilot 2006 Environmental Performance Index and the 2008 Environmental Performance Index to ensure environmental sustainability. There are various macroeconomic indicators those have got prime importance in influencing quality natural environment of a country. In this work, seven such important macroeconomic indicators viz. HDI, GDP per capita (PPP US $), population (million), life expectancy at birth (years), percentage of urban population, public expenditure on health (% of GDP) and CO_2 emission total mt. ton per year have been selected and correlation of their values for twenty-five countries including India was made with the corresponding EPI values of these countries. The countries are New Zealand, United Kingdom, Canada, Malaysia, Japan, Switzerland, Australia, Germany, United States of America, Russia, Brazil, Philippines, Thailand, Sri Lanka, South Africa, Indonesia, Nepal, Myanmar, China, Viet Nam, Uzbekistan, Cambodia, India, Bangladesh, and Pakistan.

Multiple regression model has been constructed to find out the mathematical relationship between EPI and various macroeconomic indicators of the selected countries mentioned earlier. Pearsonian correlation between EPI and each macroeconomic indicator revealed that HDI has the closest relationship with EPI whereas furthest relationship was with CO_2 emission total mt. ton per year.

INTRODUCTION

The past three decades experienced an increasing consciousness about environmental problems facing communities, nations, and the world. During this period, natural resources and environmental issues have grown in scope and urgency. Maintaining a balance between economic growth and ecosystem health has given rise to the concept of **sustainable development.** Renewable energy use, organic and low-input agriculture, and resource-conserving technologies preserve rather than degrade the environment. On a global scale, the promotion of sustainable development responds to the many resource and environmental issues in terms of ecosystem impacts rather than as individual problems. Sustainable development is a socio-ecological process considered as the fulfilment of human needs keeping the quality of the natural environment for ever. In 1972, UN Conference on the Human Environment was held in Stockholm. From then sustainability was most often talked about in terms of 'sustainable development'. In 1980, the International Union for the Conservation of Nature (IUCN) published the *World Conservation Strategy* and used the term "sustainable development."[1] In 1987, came the Brundtland Report.[2,3] The most often-quoted definition of sustainable development is coined by the Brundtland Commission. It is the development that "meets the needs of the present generation without compromising the ability of future generations to meet their own needs". Since humans depend in countless ways on the physical environment (both natural and human constructed), environmental sustainability is very much essential for social sustainability. Moreover, the viability of the economy depends on environmental resources and service flows. So, economic sustainability depends on environmental sustainability. More generally, it can be seen that environmental, social and economic policies are greatly inter-related giving rise to an intersection of sustainable policies.[4]

Fuelled by advances in information technology, data-driven decision-making has changed every corner of society, from business to biology. In the policy domain, quantitative performance metrics have modified decision-making processes in many arenas, including economics, health care, and education. The Environmental Performance Index (EPI) brings a similar data-driven, fact-based empirical approach to environmental protection and global sustainability. Policy-makers in the environmental field have begun to acknowledge the importance of incorporating analytically rigorous foundations into their decision-making although large data gaps and a lack of time-series data still hamper efforts to track many environmental issues, spot emerging problems, assess policy options, and gauge effectiveness. The EPI focuses on two main environmental objectives:

- reducing environmental stresses to human health; and
- promoting ecosystem vitality and sound natural resource management.

These broad goals also replicate the policy priorities of environmental authorities around the world and the international community's objective in adopting Goal 7 of the Millennium Development Goals (MDGs),[5] to "ensure environmental sustainability." The EPI was led by the Environmental Sustainability Index (ESI),[6,7] published between 1999 and 2005. Up till January, 2008 two EPI reports[8,9] have been prepared and released—the Pilot 2006 Environmental Performance Index and the 2008 Environmental Performance Index. There are various macroeconomic indicators those have got prime importance in influencing quality natural environment of a country. In this work, seven such important macroeconomic indicators viz. HDI, GDP per capita (PPP US$), population (million), life expectancy at birth (years), percentage of urban population, public expenditure on health (% of GDP) and CO_2 emission total mt. ton per year have been selected and correlation of these values for twenty-five countries including India was made with the corresponding EPI values of these countries. The countries are New Zealand, United Kingdom, Canada, Malaysia, Japan, Switzerland, Australia, Germany, United States of America, Russia, Brazil, Philippines, Thailand, Sri Lanka, South Africa, Indonesia, Nepal, Myanmar, China, Viet Nam, Uzbekistan, Cambodia, India, Bangladesh, and Pakistan. Further these macroeconomic indicators along with the EPI are taken into a multivariate form and Factor analysis (Principal axis factoring) has been performed with them.

METHODOLOGY

(a) Collection of Secondary Data

Different secondary data are extracted from Human Development Report, 2008[10] and EPI values of various countries are taken from EPI Report, 2006.

(b) Statistical Analysis

Bivariate analysis[11] in form of correlation and regression between EPI values and various macroeconomic indicators are performed.

Multivariate analysis was performed with the above indicators and EPI values and is expressed in terms of multiple regression equation.[12]

Factor Analysis[13] is worked out to segregate the principal components of these variables. All the analysis is carried out using the SPSS (version 12) software and MS-Excel package.

RESULTS

Correlation and Regression Analysis

Correlation between different variables (i.e. EPI and macroeconomic indicators) is depicted through a matrix (Table 1).

The regression of EPI on HDI 2005 shows that there is a positive linear relationship between them with R^2 being equal to 0.7983 (Fig. 1a). Again the regression of EPI on Log GDP per capita (PPP US $ 2005) shows that there is a positive relationship between them with R^2 being equal to 0.739 (Fig. 1b). The regression of EPI on Log population (million) 2005 shows that there is a negative relationship between them with R^2 being equal to 0.1282 (Fig. 1c). The regression of EPI on Life Expectancy at birth (years) 2005 shows that there is a positive linear relationship (r =0.375) between them with R^2 being equal to 0.5294 (Fig. 1d). The regression of EPI on % of Urban Population 2005 shows that there is a quadratic relationship between them with R^2 being equal to 0.7581 (Fig. 1e).The regression of EPI on Public Expenditure on Health (% of GDP) 2004 shows that there is a quadratic relationship between them with R^2 being equal to 0.7321 (Fig. 1f). The regression of EPI on Log CO_2 emission total mt. ton per year 2004 shows that there is a positive linear relationship between them with R^2 being equal to 0.1019 (Fig. 1g).

Multiple Regression Analysis

Multiple regression model ($y = 76.4637x_1 - 0.0002x_2 - 0.0081x_3 + 0.0375x_4 + 0.1699x_5 + 0.2738x_6 + 0.0234x_7$) has been constructed (Table 2) to find out the mathematical relationship between EPI and various macroeconomic indicators of the selected countries mentioned earlier. Pearsonian correlation between EPI and each macroeconomic indicator revealed that HDI has the closest relationship (r = +0.893) with EPI whereas furthest relationship was with CO_2 emission total mt. ton per year (r = +0.096). Further regression equations of EPI on different macroeconomic indicators was expressed which revealed the pattern of relationship of these indicators with EPI.

TABLE 1

Correlation Matrix Between EPI and Various Macroeconomic Indicators

	EPI	*HDI*	*GDP/capita*	*Population*	*LEB*	*% UP*	*PEH*	*CO_2 Em.*
EPI	1							
HDI	0.893	1						
GDP/capita	0.791	0.887	1					
Population	-0.324	-0.140	-0.175	1				
LEB	0.728	0.872	0.757	-0.078	1			
% UP	0.853	0.836	0.783	-0.184	0.6[illegible]9	1		
PEH	0.831	0.890	0.937	-0.242	0.720	0.819	1	
CO_2 Em	0.096	0.276	0.333	0.615	0.199	0.185	0.212	1

Note: EPI: Environmental Performance Index; HDI: Human Development Index; LEB: Life Expectancy at Birth; % UP: % of Urban Population; PEH: Public Expenditure on Health; CO_2 Em.: CO_2 Emission.

Fig. 1. Relationship and Regression Line Between EPI and Various Macroeconomic Indicators

Fig. 1a

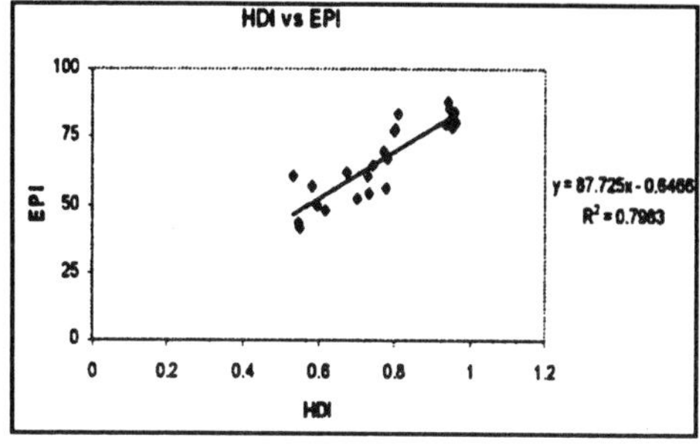

Fig. 1b

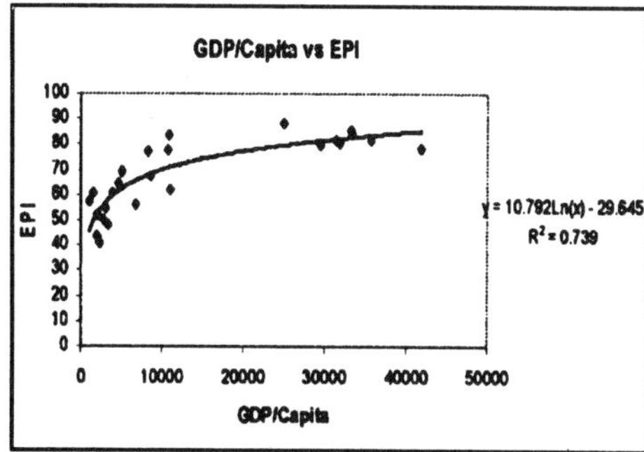

Fig. 1c

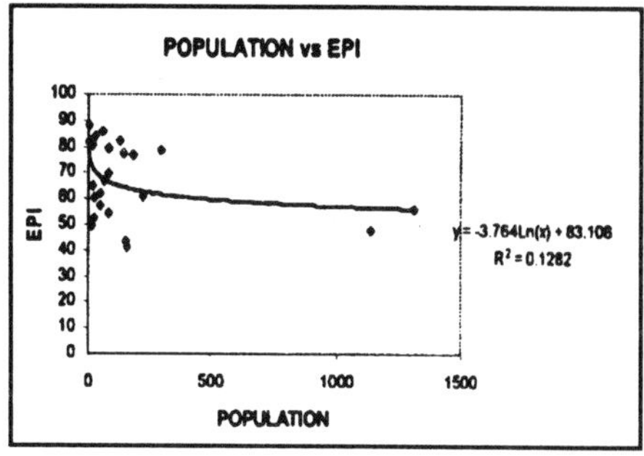

Fig. 1d

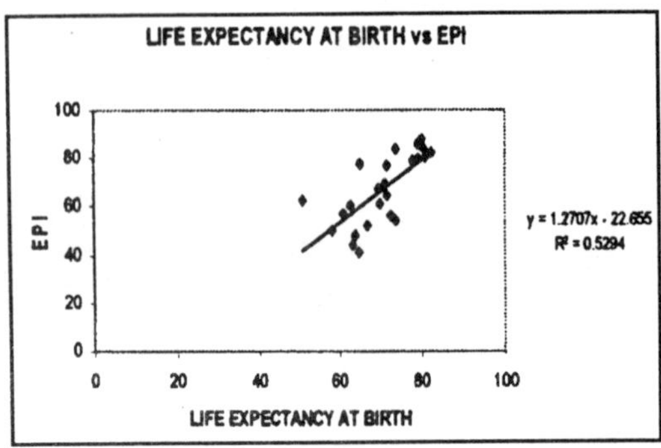

Fig. 1e

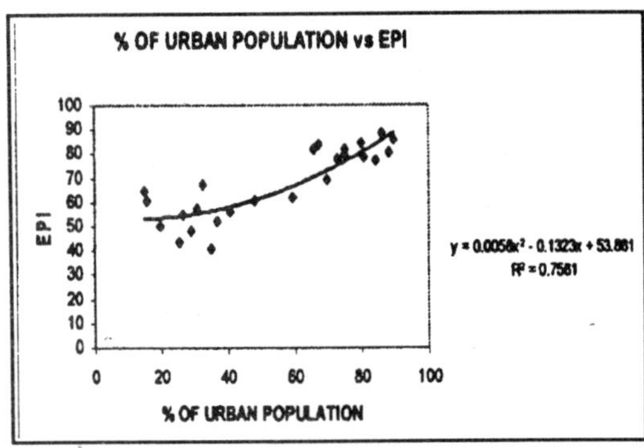

Fig. 1f

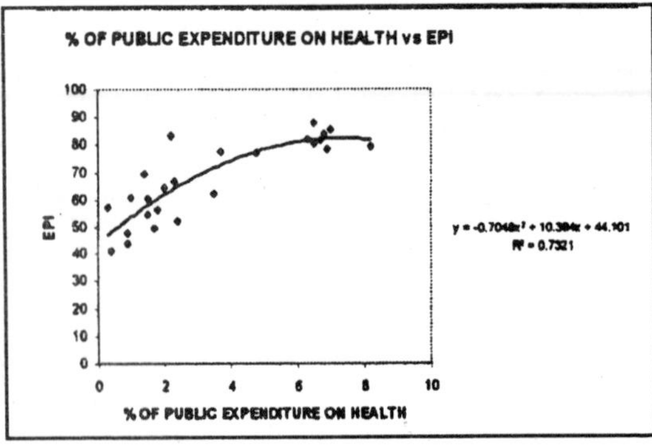

Fig. 1g

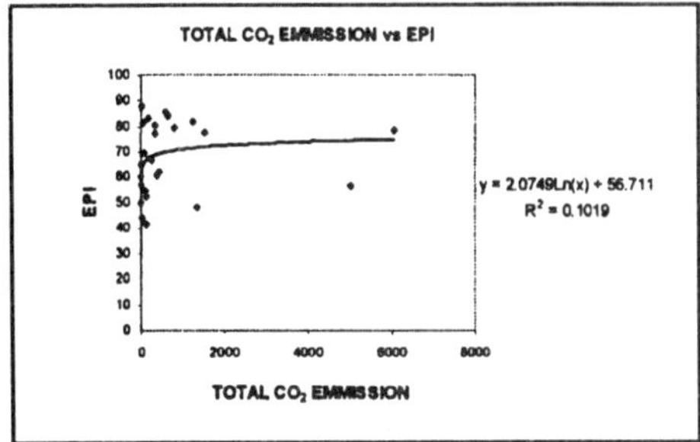

Here the result shows that the *t Stat* of each independent variable is not significant. But as a whole the *F* value is significant (Table 2 and 3).

Factor Analysis

Factor analysis is performed as a data reduction process. Principal axis factoring is used as method for Factor Analysis Extraction. Results reveal that 2 factors are extracted as revealed from Scree plot (Fig. 2) and initial Eigen values (Table 4). The distribution of variables in space is revealed through the Factor plot (Fig. 3) and Factor matrix (Table 5).

Fig. 2. Scree Plot of Different Factors

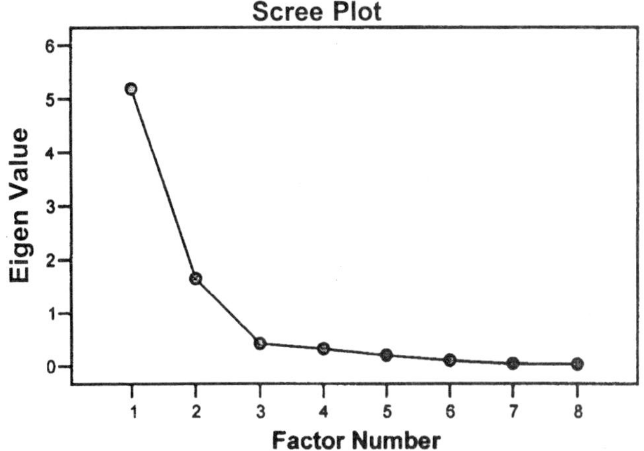

TABLE 2

Multivariate Analysis of EPI and Various Macroeconomic Parameters

	Coefficients	*Standard Error*	*t Stat*	*P-value*	*Lower 95%*	*Upper 95%*	*Lower 95.0%*	*Upper 95.0%*
HDI: x_1	76.46368	31.97966	2.39101	0.0279	9.2769	143.6504	9.2769	143.6504
GDP/capita: x_2	-0.00024	0.000284	-0.84248	0.4105	-0.0008	0.00036	-0.0008	0.0004
Population: x_3	-0.00814	0.005741	-1.41853	0.1731	-0.0202	0.0039	-0.0202	0.0039
LEB: x_4	0.037519	0.288633	0.129987	0.8980	-0.5688	0.6439	-0.5689	0.6439
% UP: x_5	0.169907	0.098059	1.732705	0.1002	-0.0361	0.3759	-0.0361	0.3759
PEH: x_6	0.273848	1.668366	0.164141	0.8714	-3.2312	3.7790	-3.2312	3.7790
CO_2 Em: x_7	9.28E-06	0.001345	0.006898	0.994572	-0.00282	0.002835	-0.00282	0.002835

TABLE 3

ANOVA Substantiating the Above Multivariate Analysis Between Different Variables

	df	SS	*MS*	*F*	*Significance F*
Regression	7	117572.2	16796.03	452.7079	4.78E-18
Residual	18	667.8226	37.10126		
Total	25	118240			

TABLE 4

Eigen Values and % Variance of Different Factors

Factor	*Initial Eigenvalues*			*Extraction Sums of Squared Loadings*		
	Total	*% of Variance*	*Cumulative %*	*Total*	*% of Variance*	*Cumulative %*
1	5.190	64.876	64.876	5.027	62.833	62.833
2	1.652	20.648	85.524	1.371	17.143	79.976
3	.426	5.323	90.847			
4	.327	4.092	94.939			
5	.205	2.567	97.506			
6	.111	1.393	98.899			
7	.050	.626	99.525			
8	.038	.475	100.00			

Extraction Method: Principal Axis Factoring.

TABLE 5

Factor Matrix

	Factor	
	1	2
y	.912	-.164
x_1	.983	.062
x_2	.932	.078
x_3	-.207	.820
x_4	.800	.061
x_5	.859	-.032
x_6	.938	-.036
x_7	.246	.810

Extraction Method: Principal Axis Factoring. 2 factors extracted. 11 iterations required.

Fig. 3. Factor Plot

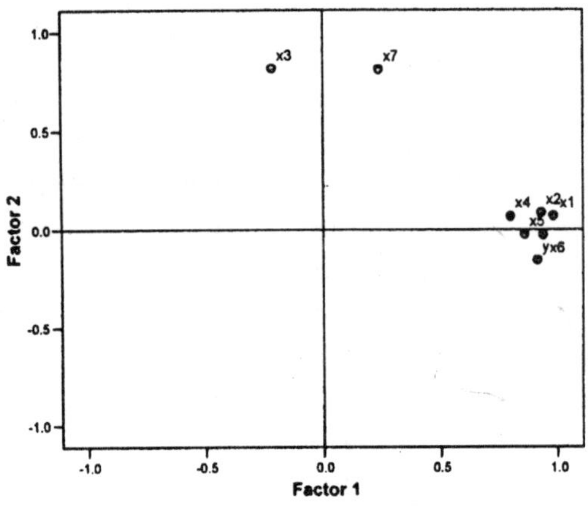

DISCUSSION

Sustainability tells about *continuity* and development about *change.* We want to sustain (maintain) as well as want to change many things. So the notion of *'sustainable development'* that combines desired change and desired continuity is quite obvious. Sometimes quite a bit of conciliation arises before a society, what should be maintained and what should be changed. But a very important task before a society arises that how to measure "sustainable development". To measure "sustainable development" several indicators have been constructed viz. Environmental Sustainability Index (ESI) and Environmental Performance Index (EPI). Aim of our work is to find relationship of EPI with various macroeconomic indicators. For this analysis we have chosen 25 different countries. According to the EPI 2006 New Zealand is the highest scorer. There are also some middle ranked countries. But we have also taken some low ranked countries like India, Bangladesh and Pakistan. EPI 2006 shows that Bangladesh, India, and Pakistan are densely populated industrializing countries with stressed ecosystems. Among 133 countries India ranks 118th, Bangladesh 125th and Pakistan 127th. Although the sample size is small the sample is representative. EPI 2006 finds out that wealth and a country's level of economic development emerge as major determinants of environmental outcomes. Every country confronts critical environmental challenges. Developed countries often experience pollution and degraded ecosystems. Every country with low EPI scores has under-invested in environmental infrastructure (drinking water and sanitation systems) and lacks the capability for aggressive pollution control or systematic natural resource management. This implies that developing countries must face the additional liability of investment in water and sanitation systems to support pollution control and natural resource management. Among many amazing facts the United States stands near the top in environmental health, but ranks near the bottom in management of productive natural resources. Moreover, densely populated countries are dispersed throughout the EPI rankings, with the highest (Japan) ranking 14th in the EPI and the lowest (Bangladesh) ranking 125th.

A pattern of relationship is drawn between EPI and certain selected macroeconomic indicators. Human development Index which reflects the progress made by a country or state in the field of Human development[14] is being correlated with EPI and it is observed that there is a highly positive correlation between the two as reflected through a linear relationship. Next, GDP/capita has a positive relationship with EPI that is countries with high GDP/capita reflect higher EPI. Countries with higher population reflected a much lower EPI value as the two variables are negatively correlated. Thus it supports the view that poor people exploits natural resources in an unsustainable way;[15] on contrary rich people destroys natural resources by increment of consumerism.[16] EPI is higher in countries with higher Life expectancy at Birth. So they have a positive linear relationship. Thus countries which have population with more longevity imply higher rank in EPI. Percentage of urban population shows a positive quadratic relationship with EPI and the regression curve shows that EPI values are higher in countries with more urban population which implies that developed countries can invest more in pollution abatement although more industrialisation is responsible for more pollution in these countries. Countries that spend more on public health are effective in scoring a higher EPI as reflected through a highly positive correlation. Total CO_2 emission seems to be unrelated to EPI value as it seems that carbon dioxide is not the lone factor in affecting the EPI of a country.

While analysing the total effect of various macroeconomic indicators and CO_2 emission on EPI values a multiple regression equation was constructed. The equation reveals that as a whole all the factors together affect the EPI values.

Factor analysis reveals that two factors were extracted through Principal axis factoring and more than 85% of the variance is explained by these two factors (Table 4). Factor plot (Fig. 3) reveals that variables, x_2, x_1, X_4 and x_5 are positively related to Factor 1 and Factor 2 whereas variable y and x_6 are related to Factor 1, x_3 is positively related to Factor 2.

In conclusion, EPI has a multivariate relationship with certain selected macroeconomic indicators depicted through a multiple regression equation. Factor analysis substantiates the principal factors and relationship of various variables with these factors.

NOTES AND REFERENCES

1. National Strategies for Sustainable Development, 2004. "World Conservation Strategy: Living Resource Conservation for Sustainable Development". www.nssd.net.
2. United Nations, 1987, "Report of the World Commission on Environment and Development." General Assembly Resolution 42/187, 11 December 1987.
3. Brundtland, G.H. (ed.), 1987, *Our Common Future*: The World Commission on Environment and Development, published as annex to General Assembly document A/42/427, Development and International Co-operation: Environment.
4. Sinha, J. and Pal, A. 2008. Economics of The Sustainable Environment. SAJOSPS, 9(1), 18-24.
5. United Nations, 2005. *United Nations Millennium Development Goals,* 2000. New York: United Nations.
6. 2005, Environmental Sustainability Index. Yale Center for Environmental Law and Policy New Haven, Conn. Yale University. www.yale.edu /esi ESI2005_Main_Report.pdf.
7. 2005, Environmental Sustainability Index (ESI), 2005, Environmental Sustainability Index. Yale Center for Environmental Law & Policy. New Haven, Conn.
8. 2006, Environmental Performance Index. Yale Center for Environmental Law and Policy Center for International Earth Science Information Network (CIESIN), World Economic Forum, Geneva, Switzerland.
9. 2008, Environmental Performance Index (EPI). 2008. Yale Center for

Environmental Law & Policy. Center for International Earth Science Information Network (CIESIN), Columbia University.www.epi.yale.edu

10. UNDP, *Human Development Report*, 2007-08.
11. Zar, J.H., 1999, *Biostatistical Analysis*, 4th Ed., Pearson Education, UK.
12. Johnston, J. and DiNardo, J., 1997, *Econometric Methods*, 4th Ed. The McGraw-Hill Companies, Inc. USA.
13. Vaatanen, P., 1980. Factor analysis of the impact of environment on microbial communities in Tvarminne Area; Southern coast of Finland. *Appl. Environment, Microbiol*, 40(1), 55-61.
14. P. Mahesh R. and Rajasenan, D., 2006, Poverty, inequality and natural resources degradation. Ph.D. Thesis. Department of Applied Economics. Cochin University of Science and Technology.
15. Roy Sovan, 2003. *Environmental Science*. 1st Ed., Publishing Syndicate, India.
16. Hamilton, K., Ruta G., and Tajibaeva, L., 2006. Capital Accumulation and Resource Depletion: A Hartwick Rule Counterfactual. *Environmental Resource Economics*, 34(4), 517-33.

Project Appraisal, Environmental Accounting and Sustainable Development

Pranab Nag and Anup Kumar Saha

The concept of sustainable development can be critically considered as a problem of dynamic optimization. It suggests for using the natural resources in a restricted way so that it is left, as much as possible, for future generation. The environmental damages that arise from the unsustainable methods of production can greatly reduce the long-term national productivity but can increase the current GNP figures. It is therefore important that the government of a country incorporates some form of project appraisal technique for the feasibility of a long term project in its development model and environmental accounting for its development planning and GNP calculation so that a real scenario of long-term sustainable growth prospect is reflected. Business community is also expected for incorporating environmental accounting in its accounts and reporting thereof to discharge its social responsibility. Environmental accounting should be a driver of the overall corporate accounting. But the business community in general is reluctant to report the detailed information on environmental impact. On the other hand, there is a reasonable ground on behalf of the employees of a business entity and the inhabitants of the locality in which the entity situates to get the information about environmental impact on the locality. The strategic use of adversary accounting can help in getting and analyzing the accounting information for the assessment of environmental impact on the employees and local community. Environmental accounting is the key to sustainable development. The present paper describes an outline of the macro and micro-aspects of environmental accounting and hints for an appropriate project appraisal model so that it ensure sustainable development instead of development. The scope of using adversary accounting in this regard is also addressed in brief.

I. INTRODUCTION

The concept of sustainable development is a recent origin to protect the environment. Though the environmental damage started since the industrial revolution, degradation of environment due to global warming, depletion of the ozone layer, water, air and noise pollution, soil infertility become the concern of the general public since the eighties of the last century. The concept of sustainable development is recognized by the World Commission on Environment and Development (popularly known as the Brundtland Commission) in 1987 to protect the environment not only for present generation but also for future generations. Sustainable development does not mean not to use the natural resources but to use these in a restricted way so that it is left for future generation as much as possible. It recognizes that environmental damage should be kept within an acceptable limit so that the nature can absorb normally within itself. Therefore a trade-off between economic growth and sustainability is necessary. The reason for exclusion of environmental cost from the determination of GNP is due to the historical absence of environmental consideration from development economics. Environmental costs can be considered in terms of social cost in the project appraisal both at the government and enterprise level.

It includes the amount of depletion of natural resources and the cost of degradation of environmental quality. Environmental damages that arise from the unsustainable methods of production can greatly reduce the long-term national productivity but it can increase the current GNP figure. The short-term economic benefit penalizes both the present and specially the future generations. It is therefore important that the government of a country incorporates some form of environmental accounting for its development planning and GNP calculation. A business firm is also expected for incorporating environmental accounting in the accounts of the firm and reporting thereof to discharge its social responsibility. A business entity is usually reluctant to report on environmental impact. That is why the conflict between the business entity and the local people including the employees may arise. Adversary accounting on behalf of the employees and local community may be useful in this regard.

Given this background, section II discusses the technique of project appraisal. Macro and micro-aspects of environmental or green accounting are discussed in sections III and IV respectively. Section V discusses the applicability of adversary accounting.

II. Technique of Project Appraisal in the Context of Environment

Project appraisal must be done before any investment to ensure a continuous flow of future consumption and environmental sustainability. It mainly consists of three kinds of appraisal, viz. (a) Financial appraisal, (b) Economic appraisal, and (c) Social appraisal. It is an intertemporal decision-making tool regarding an investment starting with identification problem followed by feasibility analysis. Feasibility analysis can be done by going through the above three kinds of appraisal.

(A) Financial Appraisal

In this appraisal we calculate net present value (NPV) of an investment. The NPV of an investment is given by the following formula:

$$\text{NPV} = \sum_{t=0}^{T} \frac{(Vt - Ct)}{(1+r)^t} - \text{K}_0$$

where,

K_0 = fixed cost the project in the base period;
Vt = value of output at time t;
Ct = operating cost at time t;
r = rate of discount; and
T = life of the project.

If, NPV > 0, the project yields a positive return.

(B) Economic Appraisal

The main concern of economic appraisal is the social cost benefit analysis of an investment. Economic valuation of an environmental benefit becomes difficult because of absence of property rights on the environmental commodity. Market price does not reflect the actual valuation of the environmental goods. Economic valuation of environmental benefits originates from three different kinds of usages of environmental goods: user value, option value and existence value. User value means the value which we have to incur to enjoy the environment directly. It is positive approach of environmental analysis. Option value follows from intertemporal and spatial difference of actual user value. It has again three components: (i) option for using the goods by the same individual, (ii) bequest value that is leaving the option open for future generations to enjoy, and (iii) vicarious value that is leaving the option open for others to enjoy. The normative approach of environmental valuation for the economic appraisal considers the existence value of the environment. Existence value means the value we want to sacrifice just because of the continuity of the environmental situation in future. As for example, we, the Indian, always agree to contribute for the campaign to protect the bio-diversity of Himalayan region not because we want to visit the Himalayan Natural Park but because the bio-diversity is required for the existence of this world. Existence value is more important for sustainable development. Therefore, considering the environmental concern, economic valuation of a project can have three different valuations viz. (a) actual user value, (b) option value, and (c) existence value.

So, total economic value = actual user value + option value + existence value

Cost is treated differently in economic appraisal method compared to the financial appraisal method. In financial appraisal, only the accounting cost is considered. But in economic appraisal, social cost and opportunity cost can be considered in addition to the accounting cost. Therefore, it can be concluded that cost is most broad based in economic appraisal than in financial appraisal.

Cost benefit analysis of environment which is the integral part of the economic appraisal process can be shown with the help of two functions viz. (i) marginal social benefit (MSB), and (ii) marginal social cost (MSC). Both the social functions are designed incorporating positive and negative externalities of an economic event. This price-based instruments are due to A.C. Pigou (*Thirlwal*, 2003). MSC of a project is obtained by adding the negative amount of externalities with the marginal private cost (MPC) of the project. Similarly, marginal social benefit (MSB) is an over estimated form of marginal private benefit (MPB) due to incorporation of positive externality.

MSC = MPC + negative externality
MSB = MPB + positive externality

The equilibrium point of socially optimum level of output of a project is determined where the MSC and MSB intersect to each other. The equilibrium situation is shown diagrammatically below:

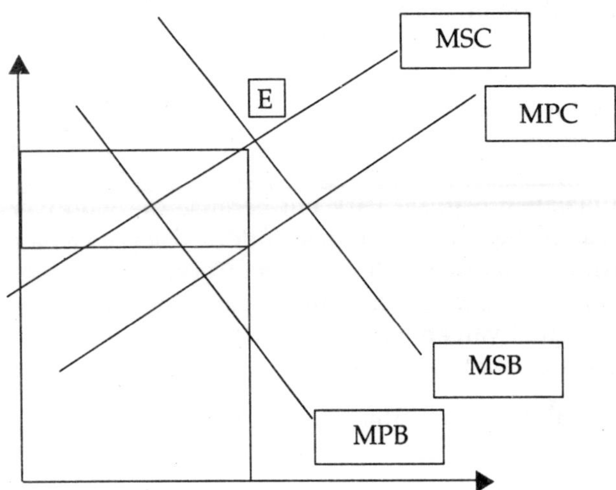

(C) Social Appraisal

Social appraisal of a project concerns with the distributional aspect. A public project may be beneficial for a group of people but may not be beneficial for another group of people. As for example, if the height of Sardar Sarovar Dam is increased then the availability of water for irrigation project and drinking purpose increases in the neighbouring states. But it certainly causes heavily to the people living in the area surrounding the dam. This appraisal is also very important in the environmental context also following the rapid urbanization program in response to globalization. Urbanization has provided benefits by the improvement of urban infrastructure but it causes environmental damages to the large urban conglomerations which affect the people who are excluded from the benefits of rapid urbanization. Therefore, consideration of redistribution is an important part of a project appraisal.

III. Macro Accounting

In conventional national accounts overall capital assets mean manufactured capital (e.g. machines, factories, roads, etc.). But environmental accounting suggests for inclusion of environmental capital (e.g. forests, soil quality, etc.) in national accounts. Sustainable development requires that the overall capital assets, i.e., manufactured as well as environmental capital assets not be decreasing and that the correct measure of sustainable national income or sustainable net national product (NNP) is the amount that can be consumed without diminishing the capital stock. It can be expressed in this way (*Todaro*, 2001:411):

$$NDP = GNP - Dm - Dn$$

where Dm is the depreciation of manufactured capital assets and Dn is the depreciation of environmental capital assets, i.e., the depletion and degradation of environmental capital assets.

Realizing the need for valuation of environmental assets and their use in national accounts, the Earth Summit of United Nations Conference on Environment and Development was held at Rio de Janeiro in 1992.[1] It was concluded in the conference that the process of development would be undermined without better stewardship of the environmental resources and services and changes therein. In 1993, the UN Statistical Commission recommended the preparation of satellite accounts (supplementary accounts) for natural resources with the core accounts. The UN-SNA (United Nations' System of National Accounts) designed a system of Integrated Environmental and Economic Accounting (SEEA). The satellite accounts relating to environment are to be prepared separately following the SEEA.

The Present SNA System

Under present SNA system, GDP or GNP is the aggregate of goods and services produced in a financial year plus (exports-imports). NDP is found out by deducting the intermediary consumption and depreciation of manufactured capital assets from GDP. Symbolically it is expressed as follows:

$$NDP - C + 1 + (X - M)$$

where

C = Final consumption after deducting intermediary consumption

I = Investment after depreciation or net capital formation (NCF)

X = Exports

M = Imports

Therefore, NCF = NDP – C – (X – M) = I.

The SNA for SEEA

Under SNA, assets are classified into two categories, i.e. produced assets and non-produced economic assets. Under SEEA, non-produced economic assets have been divided into two parts—non-produced economic assets and other non-produced environmental assets including intangibles. Produced assets include machinery, buildings, roads, live stock for breeding, agricultural crops, plantations, assets required for gases, wastes, and so on. Non-produced economic assets are those natural assets upon which ownership rights are enforced and economic benefits are accrued to their owners, for example, land, land underlying buildings, agricultural land, associated surface water, reforestation, etc. Other non-produced environmental assets include air, ocean, rivers, lakes, wild species, wild biota, forests, infertile soil and so on which are not used for economic purposes.

Under SNA system for SEEA, environmentally adjusted net domestic product (EDP) is obtained by deducting all environmental costs from net capital formation, i.e. EDP = NCF (or I)—All environmental costs. In other words, EDP is the net capital accumulation after deducting environmental costs. EDP is divided into two parts—EDP_1 and EDP_2. Environmental costs consist of (1) Depletion in produced economic assets, non-produced economic assets and non-produced environmental assets, and (2) degradation of environmental assets.

Net capital accumulation under EDP_1 = I—Depletion costs of all types of assets

Net capital accumulation under EDP_2 = EDP_1—Degradation cost of all types of assets

or, $EDP_2 = I -$ Depletion and degradation costs of all types of assets.

Obtaining NCF, EDP_1 and EDP_2 for a member of years, we can make a comparative study and find the impact of environmental deterioration on capital stock.

Valuation of Natural Assets

The valuation of natural assets is a complex one. It is mainly based on estimation. That is why controversies are there in the valuation techniques.[2] However, the UN-SNA for SEEA has prescribed some methods (*Dasgupta*, 1997). These are: (a) *Market Value Approach:* It is applied for those natural assets that have a market value in the SNA sense. For example, cultivated or non-cultivated land, (b) *Present Value Approach:* It is the sum of the expected future benefits/revenues discounted at nominal or real interested rate for the life of the asset concerned, (c) *Net Price Method:* This method is applicable for overcoming the drawbacks of present value method. Net price is the difference between the market price of one unit and factor cost per unit. The unit difference is multiplied by the total units. It can be adopted for all types of natural assets particularly for minerals, gases. (d) *Maintenance Cost Approach:* This approach is introduced in the

SEEA as an alternative to market valuation. Cost estimated or incurred to maintain or bring the environmental assets in its original form is the cost of depletion and degradation. If no maintenance is required cost of depletion and degradation would be zero, (e) *Compensation Cost Approach:* It is introduced where a community as a whole is affected. It measures the impact of environmental damage that may arise from forests clearing or land acquisition for urbanization or industrialization.

Usefulness and Limitations of Environmental Accounting

The economic growth measured by the conventional national accounting system is overstated and far from reality. Environmentally adjusted net capital accumulation and the growth rate based on these data reveal a real position of the country. It helps the government to take investment decisions and development planning for sustainable development. Environmental accounting discloses the unsound methods of production *vis-a-vis* unsound production and consumption patterns. Government may be conscious and take measures in this regard. It is useful for imposing compensation tax (pollution tax, spoiling tax, carbon tax, and the like) for internalizing the negative externalities of the polluters.

The natural resources of many less developed countries are transferred to developed countries through international trade. In the era of globalization it is now increasing in the name of free trade. Environmental accounting helps the less developed countries to understand that to what extent the developed countries are responsible for deterioration of natural resources. Most of the industrialized countries emit carbon gases (CO_2) and sulphur dioxide (SO_2) in the atmosphere, and these gases are sequestered and stored in the forests of developing countries. This pollution reduces the trees and destroys the forests of the developing countries. Thus Dasgupta (1997) observes: 'Since most of the forest land exist in developing countries, the developing countries are sequestering carbon gases of the developed countries. By keeping accounts and records, they can ask developed countries to pay for them to check deterioration. The amount would vary from place to place and region to region. The regions nearer to developed countries would be paid more. (*Dasgupta*, 1997:88). But what is happening in reality is that it has not been possible in many cases to restrict the environmental damage by internalizing the negative externalities through imposing compensation tax on the pollutant industries. Therefore, "even if the externalities are internationalized with a Pigouvian tax, economic growth will result in an increase in pollution and hence environmental decay, if the rules of Pareto optimum are adhered to" (*Khasnabis*, 2002: 4).

In European Union countries, recently there has developed the National Accounting Matrix including Environmental Accounts (NAMEA) that structures the accounts in a matrix form. It identifies the pollutant emissions by economic sector. The physical data in the NAMEA system helps to assess the growth strategies on environmental quality. If emissions are valued in monetary terms, these values help to determine the cost of avoiding environmental deterioration and to compare the cost and benefit of environmental protection.

However, the environmental accounting or green accounting is still in the stage of infancy. There may arise problems in its implementation. It is expected that gradually a common consensus on its concepts and methods will be reached.

IV. Environmental Accounting in Business Entity

Realising the need for environmental accounting and reporting at the enterprise level, the UN formed a working group known as the United Nations Inter-governmental Working Group of Experts on International Standard of Accounting and Reporting (ISAR) in 1989. The Group formulated its first guidance for accountants in 1991 in terms of what should be disclosed in financial statements to exhibit the true and fair view of the environmental performance of an enterprise. Subsequently a guidance manual was published. As per the Guidance Manual of Accounting and Financial Reporting for Environmental Costs and Liabilities of United Nations Conference on Trade and Development (*UNCTAD*, 2000), the internal costs that arise from the transactions between the business entity and another party are to be recorded. The costs which are observed and considered by the entity as external costs that should be excluded from accounting and reporting. For example, costs that may arise from the air or water pollution are not observed by the entity are considered as external costs and are not recorded in the books and reported to the stakeholders. The guidance manual again suggests that an environmental cost that is to be capitalised for recognising it as an asset should be amortized to the income statement over the current and appropriate future period (*UNCTAD*, 2000:23). An environmental liability is considered as a present obligation of the enterprise arising from past events, the settlement of which is expected to result in an outflow from the enterprise of resources embodying economic benefit.

A business entity is required to incur cost for restoration of the environmental damage (e.g. forests, fishes, etc.) and to avert the destruction of the environmental capital (e.g. air and water pollution, damages of soil fertility, etc.) that arises from the production activities and waste disposal. The costs incurred for the above can be recorded in the account books under conventional system. But the determination of the amount of environmental damage is a difficult and complex task. Because it is mainly based on estimation. That is why the estimation of cost of water and air pollution, degradation of soil is a difficult task.[3] There is a high-possibility to have a gap between the cost incurred for the environmental damage and the estimated social cost of environmental damage in the locality where the firm operates. However, survey may be conducted in the locality to estimate the amount of damage of national resources and sufferings of the people. Accounting or shadow pricing may be used where necessary to determine the cost of environmental damage. The cost of environmental damage would be (*Chowdhuri*, 1995):

$$C_{ED} = \sum_{i=1}^{n} a_i + \sum_{i=1}^{n} w_i + \sum_{i=1}^{n} n_i$$

where

a = cost of air pollution (e.g. lose of trees, agricultural crops, damage of historical building, cost of medical treatment for asthma, bronchitis, skin disease and so on.)

w = cost of water pollution (e.g. market value of insects and fishes killed, cost of clearing the polluted water, market value of crops lost for infertility of soil, cost of medical treatment for diseases arising from polluted water and so on.)

n = cost of medical treatment for diseases arising from noise pollution and so on.

The corporate responsibility of a business entity is to report the social groups both inside and outside of the firm on the environmental impacts on the society and the measures taken by the firm to restore and avert the environmental damage. If environmental costs, assets and liabilities are valued on the basis of some estimation, it should be disclosed in the financial statements.

Life Cycle Assessment (LCA)—a new methodology to assess the environmental impact has been developed recently (*Chatterjee*, 2002). The LCA covers the assessment of the environmental impact on the procurement of raw materials to be used for the products, processing of raw materials, using the product and its disposal. Coca-cola was a pioneer in the application of LCA when the company analysed the environmental impacts of beverage containers in the 1960s (*Lave*, 1996 vide *Chatterjee*, 2002: 141). A cross-country study of environmental management practices in India was recently conducted by Business Today and Tata Energy Research Institute to rate the top ten companies. The study focused also on the internal reporting system based on LCA.

At present many companies are shifting their focus away from pollution control to pollution prevention. For this purpose, a company considers clean technologies and preventive costs in capital budgeting decision. A company can invest in new manufacturing equipment that allows it to use a less costly and non-toxic direct materials. It would result savings in direct material costs and toxic waste disposal. There can also be 'hidden' impacts on a company's revenues (*Horngren, et. al*, 2003). For example, periodic training for pollution control activities is required for employees. During each training season plant remains idle though there is no idle time cost. The consequence is that training to employees adds to costs, and revenue is lost for shut down the plant for each training session.

Sustainable development requires proper environmental reporting. Sustainable environmental reporting can be ensured through the 'triple bottom line' system. It considers profit, planet and people by incorporating the economic performance, environmental performance and social/ethical performance. It is suggested by UNCTAD (*UNCTAD*, 2000: 80). The environmental performance can be reported by the environmental performance indicators (EPIs). EPI is useful for making economic and environmental decisions. Considering five global environmental problems, ISAR proposes a standardisation method for five generic EPIs at enterprise level for environmental performance reporting. The five indicators measure the five following environmental problems (*UNCTAD*, 2000: 84):

(i) Depletion of non-Renewal Energy Resources
(ii) Depletion of Fresh Water Resources
(iii) Global Warming
(iv) Depletion of Ozone layer
(v) Waste Disposal

Agenda 21 of the UNCED held in 1992 discussed some specific environmental problems that were related to three generic indicators mentioned above (i.e., II, III and V). This agenda was the most comprehensive agreement to date which was adopted by more than 178 governments participated in the conference.

In India, disclosure requirements of environmental impacts of corporates under the Companies Act are minimul. Under section 217(1)(e) of the Companies Act, 1956, as amended by the Companies (Amendment) Act, 1988, only manufacturing companies belonging to 21 industry groups are required to furnish information on conservation of energy in Form A and B as part of Director's Report. However, it would have been better to have a provision in the Act regarding the disclosure of the value of assets acquired for the prevention of the pollution or contingent liabilities that may arise from the action of the pollution control board or those of the government or any judicial forum (*Ghosh*, 2002).

V. Adversary Accounting

Accounting is generally viewed on behalf of the owners of capital. But in the context of capital-labour conflicts, accounting may be used in the interest of labour. The concept of adversary accounting (*McBarnet, et. al.*, 1993) emerges from this aspect. Under adversary accounting the financial information and accounting techniques are strategically used by the trade unions. Differing management in different issues like profit-sharing, bonus, retrenchment, an alternative approach for cost allocation, asset and liability valuation and financial statement analysis is adopted by the trade unions. An example may be given from McBarnet (1993) where adversary accounting has been applied in favour of labour:[4] "Management had consistently disclosed sales and sale prices achieved by subsidiaries in overseas markets. With the depressed domestic market, management launched a major overseas marketing campaign which involved reducing the selling prices to overseas subsidiaries and markets as a way of stimulating demand. The employee representatives were quick to realize that selling the goods at less than the standard selling price undermined their profit-sharing scheme and enhanced the profits of the subsidiaries. Challenging management on this issue resulted in a proportion of the overseas profits being brought into the profit-sharing scheme for distribution purposes" (*McBarnet, et. al.,* 1993: 92)

The concept of adversary accounting may be extended to capital-community disputes. The local people and the environmental resources of an area in which a company operates, are affected by the production and waste disposal activities of the company. There are sufficient grounds for the local people and the employees to know what are the environmental impacts of company's operation on the locality. They can ask what is the environmental policy of the organization, whether the policy is pro-people and would protect the environment properly. But the business community in general is reluctant to report the detailed information on environmental impact on the locality. Therefore, environmental information may be an issue of adversary accounting and the local civil society organizations can take initiatives in this regard. The strategic use of adversary accounting can help in getting and analysing the accounting information for the assessment of the environmental impact on the employees and local community.

NOTES AND REFERENCES

1. It may be mentioned that in the sixties and early seventies, environmental accounting was developed in Norway and France, but it

was mainly associated with quantities not with values . For a detailed discussion, see Dasgupta, N. (1997), Chapter 2.

2. For a discussion in this regard, see Sanjeevaiah, B.C. (2003), and Sengupta, A.K. (1995).
3. See Sengupta, A.K. (1995).
4. For more examples, see McBarnet, *et. al.* (1993)

REFERENCES

Chatterjee, K. (2002): 'Life Cycle Assessment: A Methodology for Measuring the 'Triple' Bottom Line of 21st Century Business', *Issues in Sustainable Development*, (ed.) R. Khasnabis, DRSP, U.G.C, New Delhi.

Chowdhuri, P.R. (1995): 'Corporate Social Responsibility Reporting: A Cost Benefit Analysis,' *Studies in Accounting Thought*, (ed.) G. Sinha, University of Calcutta.

Dasgupta, N. (1997): *Environmental Accounting*, Wheeler Publishing, New Delhi.

Ghosh, S. (2002): 'Green Accounting and Reporting in India.' *Issues in Sustainable Development, op. cit.*

Horngreen, *et. al.* (2003): *Cost Accounting—A Managerial Emphasis*, Prentice Hall of India.

Khasnabis, R. (2002): 'Jevons Paradox and The Problems of Sustainable Development,' *Issues in Sustainable Development, op. cit.*

McBarnet, D. *et. al.* (1993): 'Adversary Accounting: Strategic Uses of Financial Information by Capital and Labour', *Accounting Organisation and Society*, Vol. 18, No. 1.

Sanjeevaiah, B.C. (2003): 'Environmental Accounting', *Indian Journal of Accounting*, Vol. 34(1).

Sengupta, A.K. (1995): 'Accounting for Natural Resource Depletion and Environmental Degradation', *Studies in Account Thought*, (ed.) G. Sinha, University of Calcutta. Todaro, Michael P. (2001): *Economic Development*, Addison-Wesley.

Thirlwall, A.P. (2003): *Growth and Development*, Fifth Edition.

Swany A., Chanda R. (2004): 'Trade in Environmental Services—Opportunities and Constraints'—ICRIER-WTO Research Series No. 3.

UNCTAD (2000): *Guidance Manual of Accounting and Financial Reporting for Environmental Costs and Liabilities*, Certified Accountants Educational Trusts.

Participatory Global Environmental Governance: Mechanism for Sustainable Society

G. Sathis Kumar and S. Ramaswamy

The environmental economists approach for sustainable development, give most attention to stability of natural environmental systems, linking economic activity and environmental health. The emergence of global environmental problems such as climate change and ozone depletion has coincided with new thinking about environmental governance — do we need environmental governance at the global level? What kind of environmental governance best suits for global environmental problems? What functions of environmental are essential at the global level? Where has the existing environmental governance system fallen short? What would an effective institutional mechanism for addressing global environmental problems? The gap between environmental crises and environmental restructuring are striking force for GEG Mechanism. By considering these, the environmental economists try to buildup an appropriate restructured GEG Mechanism viz.

Three-tier Participatory Global Environmental Governance (TPGEG) Mechanism, which may provide a holistic approach for initiating the environmental protection projects for GEG.

INTRODUCTION

Emergence of global environmental issues brought the term 'environment' into the international political, economic, social, and ecological arena. The bird's-eye view pictures considerable global *environmental changes* and the worm's — eye view projects that image into serious of global *environmental crises*. These crises are translated into social conflicts, since different groups claim on a dwindling natural resource base.[1] In this context, the idea of harmonizing economic growth with the laws of society and environment, is vital for sustainable development. The environmental economists approach for sustainable development gives most attention to stability of natural environmental systems, linking economic activity and environmental health. By accepting the environmental economists' approach of sustainable development, we recognize that preservation of the environment is an integral part of global society. So, investigating the relationship between instruments of environmental regulations and economic activities is a topical issue at the national and international forums. Governments and organizations—both national and international—are facing the risk of environmental destructions, and have realized that environmental protection issues of the 21st century.[2] Present systems of production, consumption, and the governance are fundamentally out of touch with the natural world and unable to adapt changes in political, economic, social, and ecological systems. Thus, new systems of thought—production, consumption and governance—some already being developed—need to be encouraged and built upon. This will require environmental governance that is imaginative, responsive to change, and built on a clear vision to link the environment, the economy, and society. Integrating these, partnerships between States, parliamentarians, Civil society and Business are to be strengthened.[3]

With this framework, the concept of Global Environmental Governance (GEG) has been propounded and developed in and by United Nations Conference on the Human Environment (UNCHE) held in Sweden (Stockholm, 1972) which produced a declaration, action plan, and a new UN body viz. the United Nations Environment Programme (UNEP). More significantly, the UNCHE became the starting point for a large number of environmental protection initiatives at the global level. Besides, many international organizations and forums—such as G7, OECD, WTO, IMF, etc.—have put the environmental issues on their development agenda since 1970s.[4] GEG regulates public and private behaviour towards greater accountability and responsibility for the environmental protection; operates at every level ranging from the individual to the global and calls for a shared leadership and combined responsibility for maintaining environmental sustainability.[5] GEG is the sum of organizations, policy instruments, financing mechanisms, rules, procedures and norms that regulate the processes of global environmental protection. A common understanding of the term—GEG has, enabled not only governments but also various players to form norms and/or rules for solving environmental problems on their own. The emergence of global environmental problems such as climate change and ozone depletion has coincided with new thinking about environmental governance[6]—do we need environmental governance at the global level? What kind of environmental governance best suits for global environmental problems? What functions of environmental governance are essential at the global level? Where has the existing environmental governance system fallen short? What would be an effective institutional mechanism for addressing global environmental problems?[7]

GEG Mechanisms: State and Examination

Literature on environmental governance focuses on the top-down approach, according to which an improvement in environmental governance is to be sought by reforming State and State institutions, assuming that the reasons for the environmental crises are State weakness and policy failures. Hence, the process of good environmental policy-making is needed that could involve various actors of GEG like politicians, bureaucrats, public authorities, public opinion and interest groups such as environmentalists, soft technologists, environmental economists, social activists. Over the last few years a heated debate has emerged among policy-makers as well as scholars and actors of GEG on the possible and potential directions for the GEG system. The debate on GEG has focused on three ways[8]: more actors[9], more money[10], and more rules and norms."

In addition, GEG highlighting the following functions and objectives: *Integration of environment into development decision-making*—Environmental considerations should be a part of all development planning and decision-making, from the outset of policy and program priority formulation to the implementation of project-specific activities. *Trans boundary cooperation*—Sustainable regional development must recognize the trans-boundary impacts of environmental management decisions. In particular, decision-makers must be aware of both the environmental and social nature of these impacts. *Public access to information and decision-making*—Basic legal and policy frameworks should provide the public with access to information and decision-making concerning the environmental degradation that affects their lives, and with equitable access to judicial recourse in the event of damages.[12] Yet, environmental conditions are not improving across the globe and are producing a number of critical environmental problems such as shrinking forests, expanding deserts, eroding soils, deteriorating rangelands, collapsing fisheries, rising carbon dioxide levels, falling water tables, rising temperatures, more destructive storms, melting glaciers, rising sea level, drying coral reefs, and disappearing species[13]—which are producing negative environmental externalities with market failure. The hosts of problems are so long and the baggage associated with the current environmental problems is so heavy that degrade the environmental and natural resources. Thus, a fundamental restructuring and structural adjustments are needed to tinker and cement the GEG system. The gap between environmental crises and environmental restructuring are striking force for GEG Mechanism. New institutional mechanisms for better global governance are urgently needed. Accordingly, this paper examines the Global Environmental Mechanism (GEM[14]) propounded by Daniel C. Esty and Maria H. Ivanova (Yale Center for Environmental Law and Policy, Yale University) and appraising its functions and components.

OBSTACLES OF EXISTING GEG MECHANISM

The present model of GEG Mechanism talking about the three core capacities viz. *Information Mechanism* (provision of adequate information and analysis to characterize problems, track trends, and identify interests); *Policy Mechanism* (creation of a policy "space" for environmental negotiation and bargaining); and *Action Mechanism* (expansion of capacities—both global and national—for addressing issues of concern and significance). In addition, it contains the various fundamentals such as Data Collection Mechanism; Compliance Monitoring and Reporting Mechanism; Scientific Assessment and Knowledge Networking Mechanism; Bargaining and Trade-offs Mechanism; Rulemaking Mechanism for the global commons; Civil Society Participation Mechanism; Financing Mechanism; Technology Transfer Mechanism; Dispute Settlement Mechanism; and Implementation Strategies Mechanism. Through these fundamentals, the GEG Mechanism contributes to strengthen its core capacities.[15] The GEG Mechanism has many features and essential fundamentals, which moderately, benefit and assist the global governments, institutions and organizations, to frame and implement the mechanism, to manage the problems of environmental protection and environmental management. However, a close scanning of the present GEG Mechanism through the vision of the environmental economists reveals several shortcomings and they are:

- Information at different levels are not exhibited especially regional and micro-level information on the critical areas of environment are totally missing;
- Findings and inferences are not reflected in the major areas of environmental concern;
- Policy-making bodies are not properly identified at different levels. They require opinions from different experts at the global to people at the local level;
- Putting the people first and consult the stakeholders are neglected;
- Pros and Cons, Strengths and Weaknesses of policy mechanism are not clearly mentioned;
- Environmental policies are not cautiously integrated with economic, ecological, social/cultural policies of global to local and *vice versa*;

Fig. 1. Functions and Components of Existing Global Environmental Governance (GEG) Mechanism

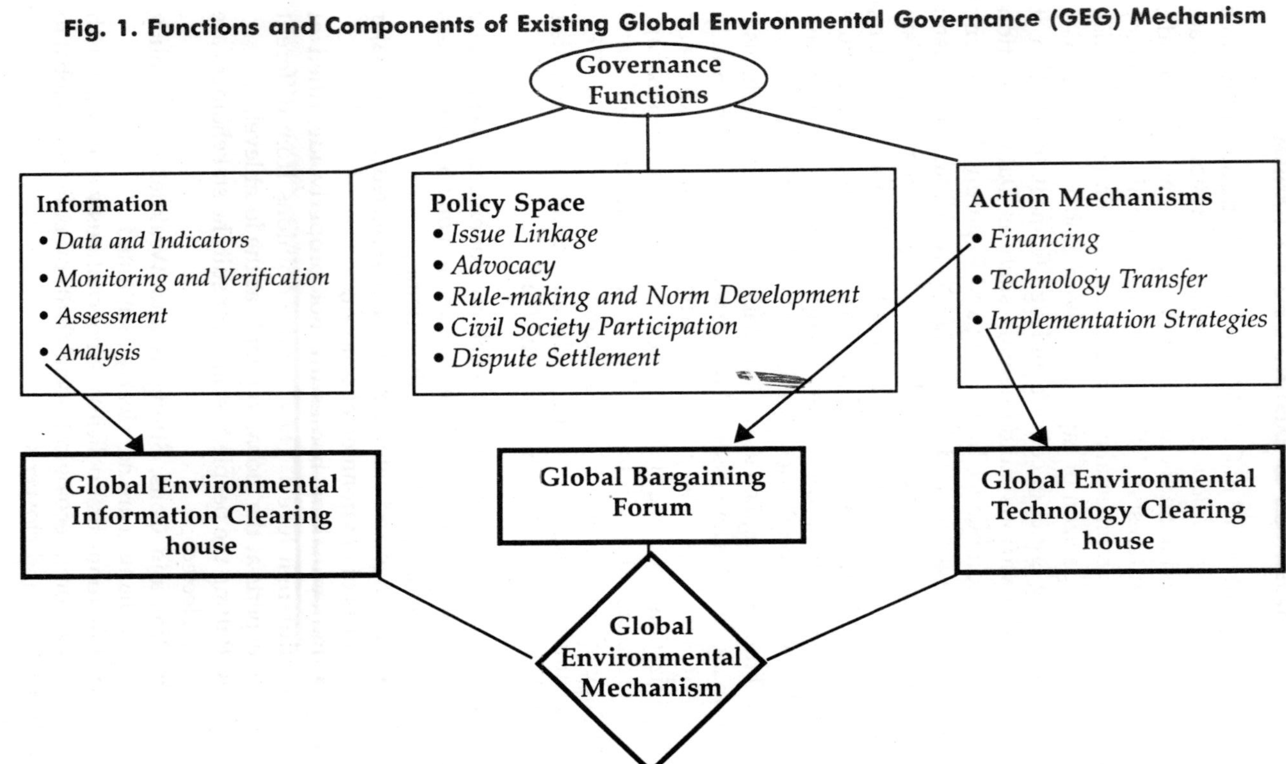

- Conflicting values and goals are not properly matching with societal goals which results in benefiting specific groups and individuals;[16]
- Though, huge funds are available for environmental protection at the global level, yet, proper monitoring mechanisms and Single Umbrella Approach (SUA) for handling and monitoring the funds are not existing in evaluating the environmental protection projects;
- Appropriate technology for third world countries is not sounded; and
- Collective mechanism and community action planning are not spelled out in implementation of environmental protection projects.

Public participation in both policy formation and decision-making, including that of indigenous peoples, access to justice, access to information, an informed, independent and unbiased judiciary, transparency and accountability, are all part of the concept of good GEG Mechanism and it must be present as well as integrated, at the local, state, regional and indeed global levels of civil society.[18] In addition, an improved quality, responsive and efficient GEG administrative regime is required to improve the GEG Mechanism which need fundamental changes in the whole GEG structure[19] through committed transformational GEG actors and players, and participation of the stakeholders in information mechanism; policy mechanism; and action mechanism at the local to global, north to south, east to west, developing to developed, haven't to have, indigenous to intelligence, and thus, bottom of the pyramid to top of the pyramid. By considering these, we—the environmental economists try to buildup an appropriate restructured GEG Mechanism viz. Three-tier Participatory Global Environmental Governance (TPGEG) Mechanism, which may provide a holistic approach for initiating the environmental protection projects for GEG.

TPGEG Mechanism: Environmental Economists Initiative

For TPGEG Mechanism, the environmental policies are to be well incorporated with economic, ecological and social/cultural policies. Besides, planning at the local to global level is to coincide with environmental planning for which Environmental Impact Assessment (EIA); Social Impact Assessment (SIA) must be integrated with Environmental Impact Statement (EIS). Environmental policy is plan or statement of intentions—either written or stated—about a course of environmental protection action, intended to accomplish in solving the environmental problems and issues at end. Environmental problems are identified and acted upon in an Environmental Policy Cycle (EPC). The different stages EPC are: Identify the environmental problem(s); Set agenda; Develop proposal(s); Built Support(s) from local to global; enact laws or rules; Implement Policies; Evaluate the result(s); and Suggest change(s). This EPC is considered as an alternative strategy to abate the environmental problems and issues, which may ultimately produce an effective GEG Mechanism. This requires a modified information systems, policy changes, and action mechanisms (Fig. 2).

Tier One—Local Players and Stakeholders

It is well known fact that information plays a vital role in framing and designing the policies, which are reflected in the development planning at the local to global level via national level. The information emanates from local reflecting the micro-level evidences and data on environmental resources and their use pattern, which exhibit the nature and intensity of environmental resource depletion and the areas of resource conservation. Local level players and stakeholders in the information system of the TPGEG Mechanism can familiar with rights and duties in providing information on environment, and natural resources management. This would help in designing micro-level policy on environmental protection and environmental projects. In order to implement the action mechanisms, grass-root level players and stakeholders are to be involved with lot of devotion and dedication. By doing so, one can bring both economic and social benefits in their regions. These benefits reflected in the national and international environmental arena by crafting good environmental governance.

Tier Two—National Players and Stakeholders

Information, policy and action mechanisms at the national level should echo the local perspectives in general and national perspectives in particular. 'Wider' and 'spread over' effects in environmental information system could reach different sections of community and institutions for generating environmental awareness and environmental safeguard from environmental degradation. Several players and stakeholders at the national level could be involved in environmental policy framework, taking into account of national environmental problems that could be the summation of local eco-economy problems. National Information System on Environment (NISE) and National Policy on Environment could coincide with each other and moreover these are two sides of the same coin. Effective implementation of environmental rules and regulations at the national level depend upon the Command and Control (CAC) policy of the government through effective implementation of legal (laws), fiscal (taxes), and economic (Incentives and Subsidies) instruments. National level players and stakeholders can act as a conciliator between local and global in the areas of environmental governance.

Tier Three—Global Players and Stakeholders

Global level players and stakeholders are to be involved in compiling and abridging the environmental information obtained from national and local players and stakeholders. This information is used as an input in framing the global environmental policy, which could be uniformly applicable to different nations of the world, irrespective of rich and poor, developed, and developing. Because, environmental concern is a global phenomenon and anything happens to environment, negative or positive, good or bad replicates at all levels and the implications are realized by all sections of the global community. Global environmental planning and collective actions advocate the international cooperation and coordination for effective implementation of global environmental policies. Global environmental players should consult the stakeholders at the local, national and global level.

Concluding Remarks

The key challenge of GEG is: how to design an institutional framework (system) that would best protect the global environment. For instance, various models are designed and endorsed by the environment players and stakeholders at the

Fig. 2: Three-tier Participatory Global Environmental Governance (TPGEG) Mechanism: Bottom of the Pyramid (BOP) Approach

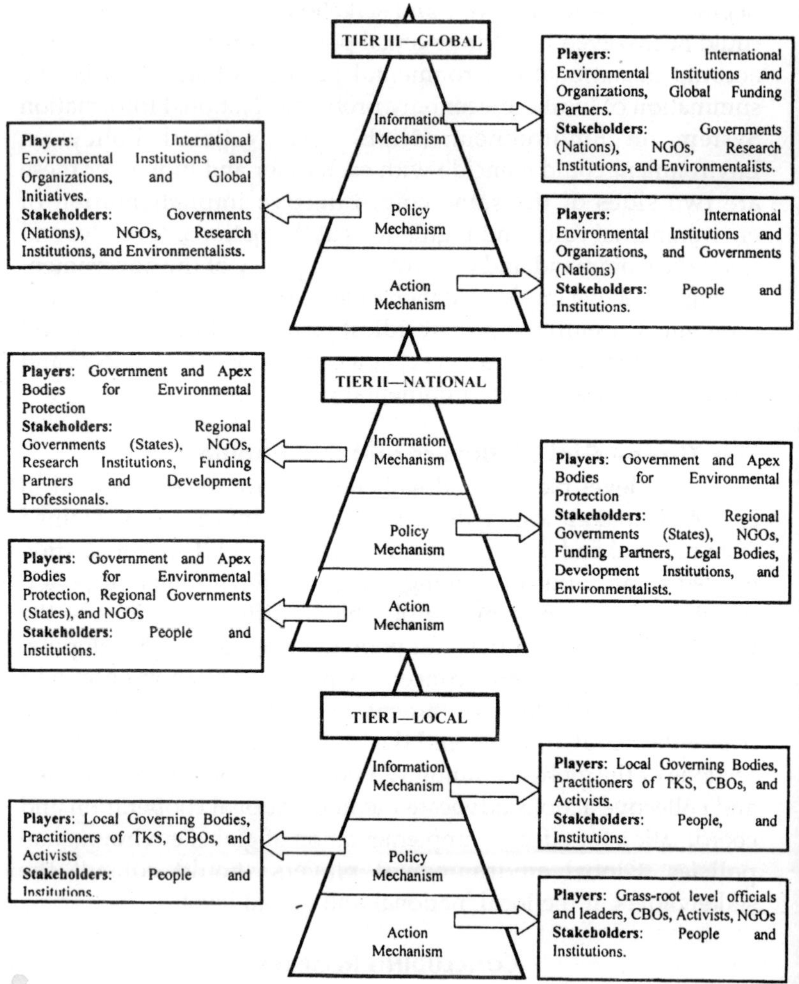

local, national, and global level. We may group the models proposed by the players and stakeholders as:

The Compliance Model: Advocates creation of a body that could provide binding decisions to hold states and private actors accountable for non-compliance with Multilateral Environmental Agreements (MEAs) and resulting environmental damage;

The New Agency Model[20]*:* Refers to creating a new organization outside UNEP with concentrated environmental responsibilities and the ability to steer and direct UN agencies in relation to environmental issues;

Upgrading UNEP Model: Takes UNEP as a departure point for improving environmental governance and suggests upgrading it to a specialized agency to strengthen its status;

Organizational Streamlining Model: Addresses the need for improved coordination and synergies among various entities within the system of global environmental governance; and

Multiple Actors Model: Argues that the system of governance comprises multiple actors whose actions need to be mutually reinforcing and better coordinated. Without better integration of these multiple actors, organizational rearrangement cannot resolve institutional problems.[21]

Any how, finally we have to depend upon the existing GEG system or any of the models endorsed and proposed by the players and stakeholders at all levels of GEG. TPGEG Mechanism does not deal to endorsing or proposing an alternative model to or add-in to existing GEG system. But, it is the mechanism, which concerns with the participation of environmental players and stakeholders at local, national, and global level on the problems and issues in environmental management and environmental protection projects. In this context, TPGEG Mechanism is proposed with an objective of improving coordination of GEG; paving the way for the elevation of environmental agenda for players and stakeholders at all levels; assisting to develop the capacity building for environmental policy; and strengthening the institutional environmental governance and mechanism for negotiation of new conventions and action programs as well as for the implementation of existing ones.

Thus, in a world affected increasingly by negative environmental impacts, environmental governance at the local to global level with an active participation of players and stakeholders becomes the key locus for innovative thinking in GEG that yields sustainable solutions for global environmental problems and issues. Experiences of the TPGEG radiate people and institutions—the prime stakeholders of the environment at all levels. Since, environment provides different inputs and performs many functions to the humanity. Environmental governance is the only way through which the humanity enjoys the environmental security and environmental justice.

Notes and References

1. Madhav Gadgil and Ramachandra Guha (2000): *The Use and Abuse of Nature*, Oxford University Press, New Delhi, pp. 2-3.
2. Skaidre Zickiene (2007): 'Cooperation in Environmental Governance—a New Tool for Environment Protection Progress,' *Engineering Economics*, No. 3 (53).
3. Domoto, Akiko (1999): 'Creating Synergy in Environmental Policy: Bridging the Gaps between the Framework,' Convention on Climate Change and the Convention on Biological Diversity. *Environmental Awareness*, Vol. 22, No. 1.
4. Bas Arts (2005): 'Non-state actors in Global Environmental Governance: New Arrangements Beyond the State', M. Koenig-Archibugi and M. Zurn (Eds.), *Global Governance*, Palgrave.
5. *Lead Global Training Session* (2006): Stakeholder Participation in Environmental Governance, Bhopal, India, 21 February—1 March.
6. Yohei Harashima (2000): Environmental Governance in Selected Asian Developing Countries, *International Review for Environmental Strategies*, Vol. 1, No. 1, pp. 193-207.
7. http://www.yale.edu/envirocenter/dated 24.12.2007.
8. Adil Najam, Mihaela Papa and Nadaa Taiyab (2006): *Global Environmental Governance: A Reform Agenda*, International Institute for Sustainable Development, Canada.
9. It includes Basel Convention, Convention on Biodiversity (CBD), Convention on International Trade in Endangered Species (CITES), Convention on Migratory Species (CMS), Economic and Social Commission for Africa (ECA), Economic Commission for Europe (ECE), Economic and Social Commission for Latin America and the Caribbean (ECLAC), Economic and Social Commission for Asia and the Pacific (ESCAP), Economic and Social Commission for West Asia (ESCWA), Food and Agriculture Organization (FAO), Global Environment Facility (GEF), International Atomic Energy Agency

(IAEA), International Civil Aviation Organization (ICAO), International Fund for Agricultural Development (IFAD), International Labour Organization (ILO), International Maritime Organization (IMO), International Strategy for Disaster Reduction (ISDR), International Trade Center (ITC), Convention to Combat Desertification (CCD), UN Conference on Trade and Development (UNCTAD), UN Department of Economic and Social Affairs/Division for Sustainable Development (UNDESA/DSD), United Nations Development Programme (UNDP), United Nations Environment Programme (UNEP), United Nations Educational, Scientific and Cultural Organization (UNESCO), United Nations Framework Convention on Climate Change (UNFCCC), United Nations Human Settlements Programme (HABITAT), United Nations Children's Fund (UNICEF), United Nations Industrial Development Organization (UNIDO), World Food Program (WFP), World Health Organization (WHO), The World Bank, World Trade Organization (WTO), and World Tourism Organization (WTO).

10. Multiple sources of funding from international organizations, UN agencies and international NGOs and the amounts are fairly large.
11. According to estimation, there are over 500 global and regional treaties and instruments relating to the environment and GEG.
12. http://www.ref-msea.org dated 27.12.2007.
13. Sathis Kumar G. and Ramaswamy, S. (2006): *Policy Interventions for Sustainable Environmental Management in India: Ecological Economics Approach.* Paper presented in the National Conference on Environmental Pollution Management and Sustainable Development, organized by Gandhigram Rural University, Tamil Nadu on 14th-15th Dec. 2006.
14. For keeping the uniformity of the terms, hereafter GEM may call as GEG Mechanism.
15. For more information refer, *ibid.*
16. Cunningham, W.P. and Cunningham, M.A. (2002): *Principles of Environmental Science: Inquiry and Applications,* Tata McGraw Hill, New Delhi.
17. Daniel C. Esty and Maria H. Ivanova: *Revitalizing Global Environmental Governance: A Function Driven Approach,* Yale Center for Environmental Law and Policy, http://www.yale.edu/gegdialogue/.
18. Donna Craig and Michael I. Jeffery, QC (2006): *Global Environmental Governance and the United Nations in the 21st Century.* Paper presented in the European Union Forum Strengthening International Environmental Governance, Sydney, Opera House, 24th November 2006.
19. *Ibid.*

20. For more details see: Steffen Bauer and Frank Biermann (2005): The Debate on a World Environment Organization: An Introduction? F. Biermann and S. Bauer (eds.) *A World Environment Organization: Solution or Threat for International Environmental Governance*? Aldershot, UK.
21. For more details see: *op. cit.* Adil Najam, Mihaela Papa and Nadaa Tāiyab (2006).

Relationship Between Environment, Health and Education: Importance of Community Participation

Sheela Datta Ghatak

We all live in and with environmental system. In Jalpaiguri town of West Bengal when we consider the size of the population in slum areas then the scale of the problem becomes more striking. In low income settlements or irregular settlements many of the people live in abject poverty and are ignored by health and educational services. The result is that the majority of urban residents have no alternative but to live illegally in self-built settlements or in dilapidated tenements. Mostly the settlements differ in some specific ways, which may be mainly depending on the nature of their populations, notably as regards migrant status, type and location of working place, income, stages in their life cycle, etc. The inhabitants in *char* areas in Jalpaiguri town are generally facing the major health problems. Their self-built settlements are likely to have a higher incidence of gastroenteric diseases due to poor or non-existent sanitation and lack of sufficient drinking-water supply including other services. Not only that housing status and high population in downtown areas lead to a higher incidence of respiratory asthma and allergic diseases, also.

Most of the people in slum areas have poor knowledge to make for a reasonable quality of life and human development. Though in these areas general people are often more exposed to flyashes, dust, noise and other pollutants but the nature of their dwellings make them less able to defend themselves against the biological and also non-biological causative hazards.

These people usually suffer from different contaminated diseases due to their traditional practices regarding food preparation, waste disposal, sanitary use and obviously they are not very conscious about their personal as well as children's hygiene.

Unhealthy environmental problems relate not only to an individual person but also to its immediate ecological factors. The basic approach adopted in the paper is to develop an equitable health care service that will meet the need of people in slums, using appropriate technologies to develop health information and awareness throughout society as a whole and also to strengthen individual and collective intersectoral responses to control over health issues. This work specifically deals with three parts of essential awareness programmes on primary health care: (a) active and genuine community participation has a major part to play in environmental health development programmes and need of proper education about prevailing health problems; (b) the prevention and control of endemic diseases in human being caused by variety of pathogenic agents; and (c) knowledge of the differences between the infectious diseases and the contagious diseases to improve the local health status.

INTRODUCTION

"Health is defined as a state of complete physical, mental and social well-being and not merely the absence of disease of infirmity."

—W.H.O.

Most of the people in slum areas have a little or poor knowledge to make for a reasonable quality of life and human development. In Jalpaiguri town West Bengal when we consider the size of the population in slum areas the scale of the problem becomes more striking, in low income settlements or irregular settlements many of the people live in abject poverty and are ignored by health and educational services. In these areas people are often more exposed to flyashes, dust, noise and other pollutants but the nature of their dwellings make them less able to defend themselves against the biological and also non-biological causative hazards.

We all live in an environmental systems. Unhealthy environmental problems relate not only to an individual persons but also to its immediate ecological factors. The stress resulting from a change in edaphic factors, produces strain within the human body to establish the physiological autonomic behavioural regulation.

The main aims of the paper are:

(i) to point out the major environmental factors affecting health in a particular low-income groups;
(ii) to evaluate the utilization by the community of existing human health and environmental conditions; and
(iii) to develop community-based resources for the upgradation of the socio-environmental situations.

The basic approach adopted here is to develop an equitable health care service that will meet the need of people in slums, using appropriate technologies to develop health information and awareness throughout society as a whole and also to strengthen individual and collective intersectoral responses to control over health issues. To improve the environmental health status of *char* area, this work specifically deals with some essential awareness programmes on primary health care, because these people usually suffering from different contaminated diseases due to their traditional practices regarding food preparations, waste disposal, sanitary use and obviously they are not very conscious about their personal as well as children's hygiene.

Mostly the low income settlements differ in some specific ways, which mainly depend on the nature of their populations, notably as regards migrant status, type and location of working place, income, stages in life cycle, etc. The inhabitants in eastern side *char* areas (those are living just beside Karala river) in Jaipaiguri town are generally facing the major health problems.

Proper public education is the main pillar for the protection of human right of life which will surely improve the environment. Health is rightly regarded as a basic human right because it increases human potentialities of all kinds.

Fig 1: Ill-Health and Host-agent Relation

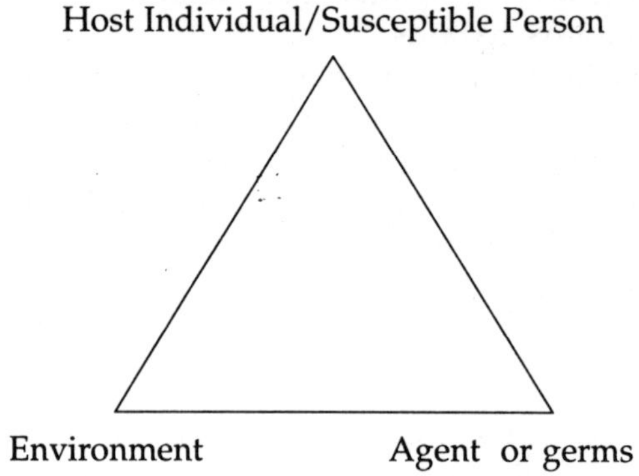

Host Individual/Susceptible Person

Environment Agent or germs

Three links in the chain transmission of communicable diseases are—

1. Reservior of infection;
2. Modes of transmission; and
3. The host.

If an environmental hazard occurs it causes damage, economic destruction, loss of human life and deterioration in health services for the community or area including flora and fauna; that means the environmental pollution also leads to degradation of plants and animals life. It occurs because production involves unwanted residuals such as smoke, poisonous chemicals and gases, noise, household wastes, etc. (Fig. 1)

Abiotic Factors and their Role in the Control of Pathogens Population Size

Now-a-days, drastic change of edaphic factors such as light, temperature, humidity, rainfall etc. may cause danger to human population by disruption of the natural ecosystem. On the other hand, regarding the rate of growth of population is combined with excessive concentration in these sub-urban areas, which points to the backward structure of an economy. Generally the level of literacy of an area determines the quality of population of that particular region and thereby it effects the advancement of that locality, though the spread of education in different sectors has not been of a uniform nature.

Here the economic condition and living status of the people in char-slum areas (CSA) in Jaipaiguri town are displayed in Table 1 on the basis of a sample survey of 100 households.

TABLE 1

Household Income of CSA per Month (approx.)

Income range (Rs.)	*Percentages of households*
450- 600	40
600-1000	32
1000 3500	18
3500 and above	10
Total	100

In this CSA the children are mostly malnourished and more than 70% of the family live in a non-ventilated one roomed unit; normally they live in huts. During survey, it could be noticed that houses without proper ventilation and also improper maintenance of household materials result in the formation of unhygienic microclimate which are very much comfortable and ideal breeding corner as well as habitat for different pathogens and their carrier also. Often various common contaminated infectious diseases also may break out and in many cases they prove fatal.

Among various macro and micro-associates of house dust mite *Dermatophagoides farinae* and *Dermatophagoides pteronyssinus* are known to be the most potential energy causing organism and prevalent in different houses in India (including Jaipaiguri) and also throughout the world. This paper presents an account of the occurrence of a large number of mites belonging to the genus *Dermatophagiodes* sp. (Pyroglyphid mite), isolated from the collected house dust samples, which sometimes cause irritation of the respiratory tract, leading to adverse effects on health. These are found to occur throughout the year indicating the capacity to withstand the extreme climatic conditions of Jalpaiguri town. However, they were found to attain their peak populations at different times of the year, i.e. mainly during the month June to July for this allergy causing mites. (Fig. 1A)

On the other hand, in these poor settlements practices of using water from polluted puddles, ponds and Karala river for all households and cleaning purposes by the slum-dwellers have rendered increasing risk for spread of different chronic diseases and sufferings.

Polluted water actually harbour or a nice shelter of pathogens of cholera, typhoid, diarrhea, jaundice, hepatitis as well as eggs and larvae of a number of parasites like tape worms, round worms, hook worms, some protozoans and micro-arthropods. [Fig. 1B(a-d)].

On the other hand, in slum areas, low income people have very low access to safe domestic water and adequate sanitation. There are some most common environmental scenario, such as foul smell, blocked drains, mosquito biting, water logging and flooding of dirty water mainly during rainy seasons. Because of non-availability of clean and safe drinking water and other domestic uses poor people are compelled to drink polluted water and to have bathing and washing which adversely affect the human health. Inadequate sanitation facilities forces the slum-dwellers to dispose solid and liquid excreta and wastes in and around their settlements, which cause harmful effects to the immediate surrounding as well as the larger environment, where both infection and contagious or contaminative diseases may flare up in alarming proportion because of these unwanted environmental pollutions. (Fig. 2A, 2B, 2C)

Different Remedial Measures

Due to different housing problems some important remedial measures have to be taken or adopted mainly for the management of disturbed environmental conditions through proper education.

(a) Better safety measures are to be adopted for working

1A: House dust allergy causing astigmatid mite (Dermatophagoides sp.).

1B: Pathogenesis of some protozoan parasites (a-d).

(a) *Plasmodium vivax*—Malarial parasite.

(b) *Giardia intestinalis*—Diarrhoeal parasite causes Giardiasis.

(c) *Balantidium coli*—Causes diarrhea and dysentery.

(d) Trophozoite (active pathogenic form) of *Entamoeba histolytica*—Produces amoebic dysentery.

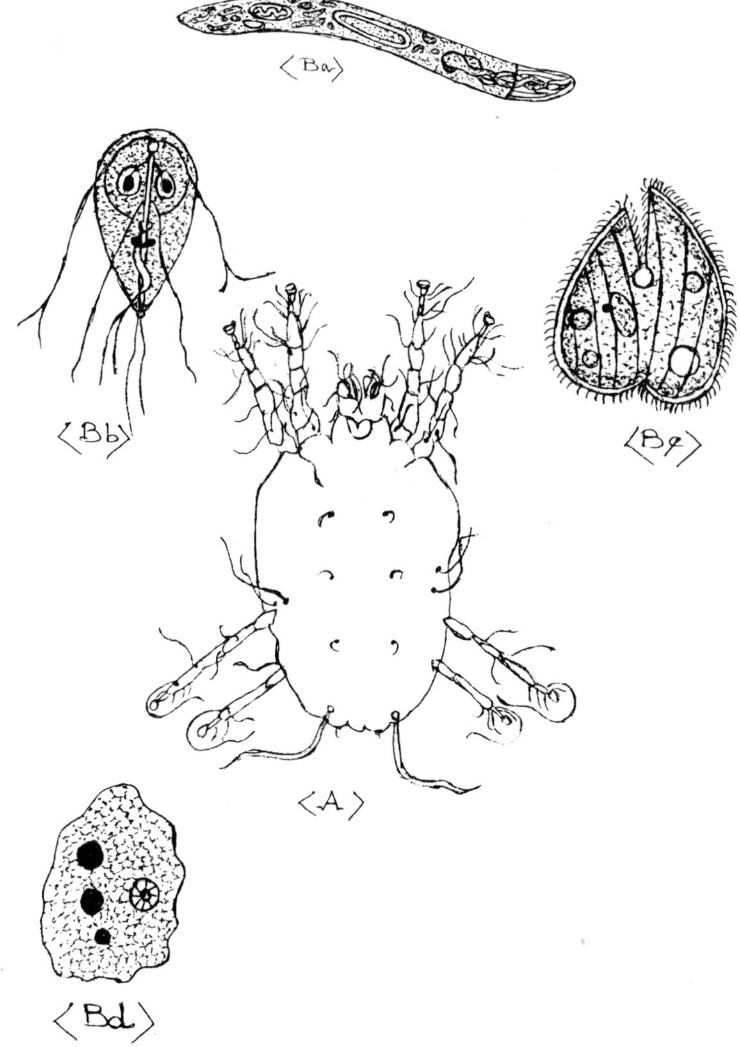

Unhealthy Environmental Problems cause Health Hazards

Fig. 2A: Vectors and Pathogens Cause Human Health Hazards

Fig. 2B: Water-borne Diseases are Directly or Indirectly Related to Contaminated or Polluted Water.

Fig. 2C: Collecting drinking water from Contaminated Water Reservoir in Dirty Place.

persons handling waste materials. In squalid areas human health hazards may be faced by direct contact, that is, in case of workers handling dirty wastes and the rag pickers collecting recyclable materials. These persons directly may acquire pathogens through contact on the skin or through inhalation. These people primarily use to dump these collected dirty materials in the surroundings of their dwellings and generally suffer from skin itching, respiratory asthma and allergies, tuberculosis, gastrointestinal disorders.

(b) Before disposal at first the biodegradable and non-biodegradable wastes are to be separated.

(c) Organic wastes are to be disposed in suitable sites, far away from living areas, to prevent spread of infectious causative agents. Actually organic wastes including human and animal excreta as well as sewage materials are collectively called "Oxygen demanding waste." The oxygen demand develops due to heavy growth of aerobic bacteria which use-up oxygen while decomposing the waste.

(d) Special importance or care and preventive measures should be taken against reusing and for safe disposal. Disposal of plastic waste management becomes a difficult task which needs community awareness. People should know that plastic as polymers are not harmful to human body, because these are inert and non-toxic but the monomer units of some polymers (plastics) are terrific harmful and dangerous to human health. Carelessly throwing of plastic bags and other such products everywhere may cause eyesore; often these wastes find their way into the drains, sewers and block them, which becomes a serious threat particularly in urban areas. Actually non-biodegradable plastics remain in soil and garbage dumps for about thousand years. As a result, soil porosity and fertility being lost, water-bodies become polluted and blockage of drainage and sewers causes distress particularly during rainy season basically which is normal peak breeding season of mosquitoes and which has become a serious threat to human health.

(e) Diseases can be prevented from entering human's body

by educating people on proper sanitary laws and hygiene. Solid wastes contain human faeces with enteric pathogens or causative agents of dysenteric condition such as *Shigella dysenteric* which can be transmitted by different insect vectors or a protozoan parasite *Entamoeba histolytica*. Good sanitation facility along with individuals and community education on cleanliness and hygiene along with consumption of fresh food and disinfected water are the keys to prevent spread of pathogens or risk factors and control of disease for keeping good health as well as clean environment.

The women should also come forward with a constructive attitude through their supportive activities based on wide co-operation, mutual understanding with sincere respect that will help to form a healthy community structure. It is heartening to express that in every year on eighth day of March, the "International Women's Day" is being celebrated, but reality is different from what the celebration means to suggest. Unfortunately, yet women do not get actual respect to take active part particularly in decision-making policy which is comparatively high in low developed areas like slums. With this another important factor is the level of education of mother—to prevent child mortality and also to educate children through their mother, because she is the best teacher, guide, friend and dearest person to a child in a family.

Practically, educational level itself plays a direct role in reducing birth rates. It is a vital point that education of both male and female affects fertility by influencing the knowledge for having small-size families and simultaneously by increasing the age of marriage mainly for the girl child. Because child mortality seriously depends upon a number of complex and interrelated set of factors such as biological, social, cultural, and economical status. With the increase in the education level of mother, the child mortality case is rare, whereas, the child mortality rate is very high among illiterate mother in slums. Hence, it could be observed that the level of education of each and every mother has a very strong impact on lessening the child mortality.

Though there are several reasons and many examples for the rapid growth of population in the slums. Main important reason is the poverty of the people. Greater the poverty—lack ol education —larger the size of the family. Illiteracy among the people is another factor. A number of parents have a strange attitude that if they have more children they can earn more money by them; also most of the people in slums are illiterate and poor as they can not afford the cost of education of their children.

Another serious issue is child laborer, the main source of which is different slums. This system is not new at all, it was prevalent even in the old days. Though with a passage of time and with the rapid population growth, it has become a social as well as an environmental problem. In fact, children working in industries and also in different non-industrial occupations are dangerously affected retarding their physical, social, moral and moreover mental development. Because of so many reasons innumerable children, both girls and boys in their lower age are compelled to come out of their families in search of work and people make themselves available for hard job, when they get seriously ill and carry some unwanted diseases. Thus now-a-days, the problem of such labourers is an alarming issue all over the world.

CONCLUSION

The approach adopted here seeks to encourage change through involvement of the community as a whole to the initiation and sustainment of the process of social and environmental improvement. Sometimes in a decision-making process, participation becomes very nominal and does not actively involve the total community, whose outcome may affect the residents and environment in a non-optimum manner. In fact, it is increasingly being recognized that community participation has a major part to play in development programmes and offers several distinct benefits. Specifically, community participation leads to a direct match between the actual need and what they eventually get, what to do and what not to do. Moreover, an active part in decision-making generates greater satisfaction with the final result and a greater commitment to use and proper maintenance of product. Active and genuine community involvement is prerequisite for identifying environmental health problems, determining their order of priority and implementing measures to deal with them. In fact, within a community there are likely to be many different types of participants or community residents including aged persons, midwives, housewives and other residents with poor or lack of knowledge, etc.

People should be made more cautious about health and the risk factors in their surroundings. They also should be made aware of their civic sense. Environment-damaging activities should be banned and in this respect help of laws, i.e. human rights, human health rights, women's rights, child labour rights, etc. and social welfare acts should be known. If necessary amendments to rules (it any) can also be applied with proper notification to the slum-dwellers as well as to the general people. They also should remain conscious that violation of these rules should be heavily penalized.

Finally, it can be stated that "live and let live," policy should be our motto and by remembering this ideal principle we should do everything good for the well-being of ourselves and to save properly our motherland.

References

Datta, S. and V.C. Joy, 1987, Incidence of a Potential Allergy Causing House Dust Mite, *Dermatophagoides farinae*, Hughes, at Santiniketan. First National Seminar on Acarology (abst) 38.

Datta, S. 1998, Asthma and Allergies: A Cursory Glance; *Frontier News*, 6, Nov. 15.

Dutta, S. 1998, Seasonal dynamics of allergy causing house dust mites, *Dermatophagoides* spp., in homes at Jalpaiguri, Fifth W.B. State Science Congress.

Datta Ghatak, S., 2000, The unhygienic life style of the migrated people living in urban-slums: A heaven for allergy causing dust mites. Urbanization Migration and Emerging Problems in Indian Situation, April 22-24.

Saxena, R.K. *et al.* 1979, Role of mites in House dust allergy. First all Indian Symposium, in Acarology, 1979.

W.H.O. Task Ground, 1990, Potential Health Effects of Climatic Change. Geneva.

SECTION III

ENVIRONMENT AND SUSTAINABLE DEVELOPMENT: SPECIFIC ISSUES

THE RELEVANCE OF 'INDIGENOUS PEOPLES'

A CASE STUDY OF THE RAJBANSI COMMUNITY OF NORTH BENGAL

ASOK DASGUPTA

Application of the universally accepted dogma of 'Indigenous Peoples' upon all the aborigine folk peoples scattered world-wide asks for systematic approach. Such a given community situated within a specific geo-ecological region of northern West Bengal, India for long could be subjected for postulation of the approach. Rajbansis here have transformed from a simple community to a complex social fold and adapted certain local characteristics that could promote their separate identity again with certain regional variations. They in the age-old transnational trade route of North Bengal have admixed with other peoples and saturated with their cultures and *vice versa*. In this way, a marked difference could be developed from their fellow peoples living in other parts of Bengal (the entire West Bengal continuous with independent countries like Bangladesh and *Moran*, the eastern foothills of Nepal and adjacent areas like eastern part of Bihar state, Assam state and Tripura state). These people are chiefly agriculturists, know well about the ancient river routes passed through dense forests and also belong to caste hierarchy ascribed within Hinduism and occupy a prestigious rank of Bratya-Kshattirya with definite traces of Buddhism, Hinduism (both caste-based close-ended as well as egalitarian versions), Animism, Syncretism, prayer to snakes and rivers, worship of Fertility Cults and local agricultural deities plus regional versions of Islamic heritage. No doubt they have developed certain values and beliefs quite non-functional in attitude and symbol-oriented in nature. The approach towards investigating the applicability of the label 'Indigenous Peoples' upon a given population demands an in-depth observation within indigenousness and also the cultural lag created due to certain external influences (modernity, globalization and anti-globalization, migrations and culture contact, multiculturalism, politico-economic impetus, monetary economy and so on).

A **three-level-approach** could be used as the methodology for the justification: firstly, attachment with land and historicity; secondly, role of Indigenous Knowledge System, and lastly, the effect of globalization (most viable way of external impact) upon IKS dedicated to certain global public and environmental services.

Here, the Rajbansi Social Fold within the multicultural scenario of North Bengal has been taken as the reference for this very discussion.

INTRODUCTION

Traditional Knowledge System (TKS) is the set of knowledge traits produced through day-to-day life experience via trial and error; these traits are asymmetrically distributed worldwide among various gender and age groups, socio-economic strata as well as ethnic-identity holders— each residing in a particular eco-geographic habitat provided with specific bio-diversity and gene pool. A TKS has two basic services: protection to Nature by virtue of proper balance between mode of exploitation of ecosystem but with certain amount of feed back and the minimum energy requirement of a given-size population attached to its folk life. Now, the supporters of Global Market Economy are of opinion to create a universally applicable "Indigenous Knowledge System" (IKS). To them, this system would behave as a part of the Global Knowledge System and simultaneously work parallel to the Global Market Economy on capitalism. It would deliver some sorts of Global Public Service (GPS) through summation of all the TKS and therefore help in reducing the negative impacts of Global Market Economy (such as pollution, biodiversity-loss, ecological degradation, emergence of new-drug resistant disease strains, genetically modified food, malnutrition, increasing levels of non-degradable and toxic compounds, global warming, drastic need of market expansion, severe competition and consumerism, outsourcing and unemployment, economic inequality and crisis-prone financial market, one-way development, mass-exploitation of nature, development of the system of production-consumption-governance but without any proper way of distribution, conflict between globalization and anti-globalization, localization-on-terrorism, fuel-crisis, war-on-resource-and-energy, space research and so forth). Besides new technological advancement; a balance between the Global Market Economy and the newly fashioned IKS is highly needed for two primary reasons: protection of nature through sustainable way of development and secondly, to check the least scope of emergence/reoccurrence/over-dominance of any singular alternative. The alternatives might be on the line of Socialism, various types of Extremism and/or any of the traditional politico-economic systems. Each of them has certain negative and positive roles in society and the Global Market System often makes propaganda of them. In favour of construction of IKS, all the folk communities (each designated as indigenous community) have now to be gathered under the global category of Indigenous Peoples and their respective cultures as Indigenous Culture. This is another kind of Globalization to save the present Globalization of the West who believes in regulation of nature on technology-and-science and no Super-Nature. Backward communities—the precious suppliers of the Global Public Service—are traditionally highly against the Western way of Globalization and therefore they have to be brought under confidence via the elite and advanced sections among them. They have to be provided with universally applicable protective measures; ILO in 1991 has proposed Indigenous Rights for all the Indigenous Peoples. Indigenous Rights are further sub-categorized into the domains like general policy, land, recruitment and conditions of employment, vocational training, handicrafts and rural industries, social security and health, education and means of communication, contracts and co-operations across borders, administration, general provisions. In a more formal side, these rights are also related with other issues such as prevention of bio-piracy, check to illegal knowledge-and-technology transfer, and performance of sustainable development, protection of basic human rights, rights for the minority communities and weaker sections, intellectual property rights as well as suitable development in association with natural resource management, human resource management, knowledge management and so on) (Fig. 1).

There are again two distinct problems:

1. New way of colonialism

(a) Scope of interference into the internal matter of a country and therefore, deviating that country's Constitution, administration, state machinery and Legal Provisions

(b) Conversion of culture, society, knowledge, intellect, instruction, human resource and natural resource into capital

2. Scope of ethnic conflict among various folk communities regarding their degree of indigenousness

Three-Level Approach

In this paper, relevance of the term Indigenous Peoples is going to be checked out for the Rajbansi Social Fold of North Bengal, West Bengal, India through a three-level approach. These three approaches are *attachment with land, role of IKS and response of Globalization.* For a community living in rural areas and sharing the attributes of folk life, is it possible to mention it under the category of indigenous peoples even when it has been actually passing through the process of modernization and started loosing its IKS. Importance of this three-way approach is going to be discussed here in brief.

Rajbansis constitute the backbone of the agrarian social structure on the planes of North Bengal. They have been

Fig. 1: Relationship of Indigenous Rights with other fields like Technology (A), Knowledge and Bio-Piracy (B). Property and Rights (C), Various forms of Capital (D). Sustainable Development (E). Folk Life and Indigenous Peoples (F) Bio-Diversity & Life Management (G), and Globalization (H)

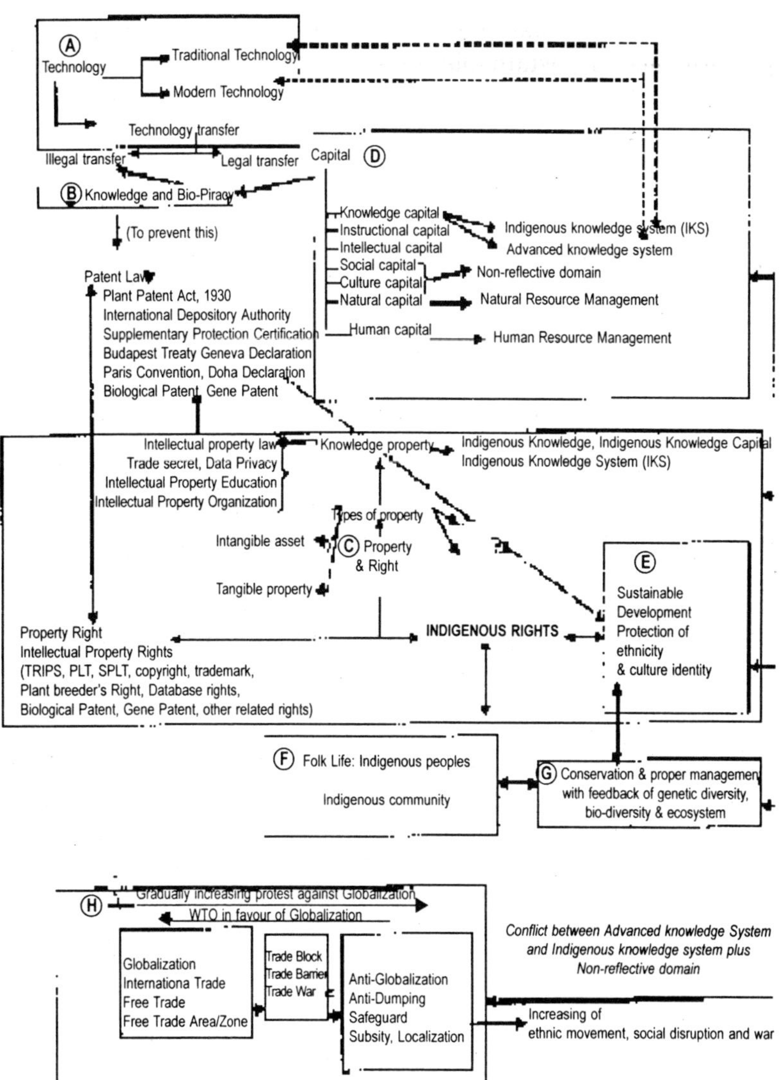

thoroughly nourished by these animists, Buddhist, Hindu, Islamic as well as Western connectivity; and could also claim their belongingness to the ancient state of Paundrawardhana of undivided Bengal. According to Sanyal (1965), these rulers typically of Kashyapa-Bratya Kshattriya combination have now turned down to the status of simply agriculturists and in North Bengal and its adjoining areas and developed themselves as the Rajbansis—a unique Social Fold. They should be able to constitute a very important IKS applicable in service of the bio-diversity, sustainable development and community welfare of North Bengal.

Level I: Attachment with the Land and Indigenous Peoples

Ideally, if the community is in touch with the land for long so far to be recognized it the aboriginal or assimilated with the aboriginal, it may be considered as Indigenous Peoples, at least for its traditional rights on the land according to the indigenous rights prescribed by ILO. But these autochthons may not be deprived and part of the mainstream. They may be also in the process of this mainstreaming.

Pre-Colonial Perspective of Indian Society (with special reference to Eastern Part of Indian Sub-Continent including North Bengal)

Trans-National trade [between Tibeto-Burmese belt and China on one hand and South-east Asia, eastern coast of Indian Peninsula, island of Sri Lanka and Kerala as the gateway of Orient for the West and Arab countries on the other] is the most crucial aspect in this context. Trade through the high mountainous passes of Sikkim, Bhutan and NEFA, then the dense forest in the foothills of Duars/Doors and the river ways of the fertile plains of North Bengal once brought prosperity to entire agrarian Bengal. Mixed economy of trade and peasantry on folk life was quite responsible for formation of several small states or settlements in North Bengal highly continuous with eastern part of Undivided Bengal landscape. These include:

- Worshippers of various fertility cults, mother goddesses, Shiva, snakes, birds, ghosts, angles;
- Kirata the hunter-gatherer and fishermen;

- Paundra-Kshattriyas (rulers-*cum*-cultivators of ancient Paundrawardhana state at the core of entire eastern Sub-Continent);
- Kashyapa clan holders (very old inhabitants throughout the entire agrarian plains of Bengal);
- Aryan/pro-Aryan stalks;
- Buddhist fragments;
- Pro-Kushana opening the trans-national trade in name of Buddhism throughout the so-formed Buddhist World;
- Barmana Kings of eastern Bengal and Assam;
- Pala Empire on trade and cultivation throughout the eastern Sub-Continent for 400-year, Kamboja-Pala rulers connecting the trade zones of Assam-Bengal and Kashmir-Punjab-Kabul-Central Asia;
- Kaivartha-Mech the fishermen-*cum*-regulators of river routes of North Bengal;
- King Jalpa the symbol of assemblage of pre- and post-Aryan Hinduism with Buddhism and convergence of Kirata elements/Paundra-Kshattriya/Bratya-Kshattriya the excluded rulers from Hindu caste hierarchy/ Kashyapa clan-holders/Kushana and pro-Kushana/ Barmana/Kamboja-Pala so as to form Rajbansi social structure (at Jalpesh/Jalpaiguri);
- Khen the oil extractor-*cum*-traders (at Kamtapur);
- Tribesmen from Tibeto-Burmese belt: Garo, Bodo, Kuchhur, Koch (at Coochbihar and Baikunthopur), Mech, Rabha, Lepcha (now in Sikkim) from Burmese-Naga region as well as Bhutia/Dukpa (in Bhutan)/Denzongpa (in Sikkim)/Sherpa (in Nepal)/Monpa from southern Tibet; some Mog-Arakan fragments regulating the major river ways and costs of Bengal;
- Aborigines like Toto, Doya, Tharu, Dhimal, and other abolished categories (Jalda, Mon, etc.);
- Pure Hindu elements-Brahmin, Kshattriya, Vaishnava, Nath/Debnath and various Bengali Hindu caste groups;
- Turk-Afghan descendent from Central Asia and Afghanistan-Punjab;
- Sufism;
- Semi-independent autonomous states; and
- Coochbihar under the Koch-Rajbansis during Mogul-

Rajput coalition on the notion of One-India until the British colonizers came in power.

This historicity is a proof of closeness of North Bengal (and the entire Bengal) with Assam, Manipur, Surma valley and Arakan coast of Burma/Bagan (today's Myanmar). But what is more important to remember is that every time there were the agriculturist communities on the fertile planes. In case of North Bengal, the major agriculturists were today's Rajbansi people, categorized by the northern branch of the ancient ruler-*cum*-cultivator Paundra-Kshattriyas/Bratya-Kshattriyas/Borgo-Kshattriyas. They could be identified by the common clan Kashyapa as associated with the myth of Parasurama in Hindu Puranas.

From hoary past, competition and association between trade and agriculture were always there in Indian Sub-Continent:

1. Between worshippers of fish-tortoise-snake and bird worshippers, Danavas and Devas, Rakshashas and Aryas, Ravana and Rama;
2. Pure Arians stalks and other non-Arian, founders of Indus valley Civilization and Arian stalks;
3. Pure and Intermixed, Brahminical and non-Brahminical heritage, Brahmin Parasurama and Bratya-Kshattriyas, quasi-matrilineal Pandava and strict-patrilineal Kaurava;
4. Semi-nomadic tribes from Silk Route in Central Asia/Iran/Tibeto-Burmese plateau and followers of Indian Agrarian Civilization, concept of Arya and Buddhist Philosophy, pre-Arian Hindu stalks like Chola and Buddhist-Arab combination and so on.

Colonial Perspective of Indian Society (with special reference to Eastern Part of Indian Sub-Continent including North Bengal)

From Nepal, Rajmahal and Central India; the British brought in Nepalese ethnic groups under the Gurkha regiment in 19th century British India as well as Santhal on agricultural sector and other Adivasis in tea estates so formed to challenge against the Tea Economy of China (i.e., the heartland of so called Buddhist World). Nepal was then governed by the Gorkha House of

Gorkha-Katmandu region with an idea of Pan-Himalayan Statehood formation irrespective of Buddhist and Islamic impetus; the policy was quite successful where there was no major trade pass from Tibet, i.e., within the region from Shipki la/Shimla to Jelep la/Sikkim-Bhutan-Jalpaiguri and Coochbihar. That was the first time when Nepal was under the hill people than the old dynasties of Terai foothills. After the Muslim invasion in North India in 10th-11th century; the Rajput stalks with their notion of racism (concepts of Arian) and idea of Hindu cultural super-ordination (concept of Arya) were shifted to the tribal states of Central India and remote areas in Himalayas, especially the region of Gaharwal Hills, adjacent to western Nepal. Probably in order to reduce the Buddhist and Muslim impact on the trans-national trade in North Bengal, British took the policy to create a demographic shift by introducing Nepali ethnic people and Adivasis there and develop a new economy on a Multicultural situation. That was highly crucial to establish control over the buffer pockets of Coochbihar State at northern margin of Bengal Presidency. That was the same economic policy that the British East India Company had to deal with other European Companies and Indian Trade Houses during establishment of their direct and indirect control over the dumping grounds and port areas like Bengal, Andhra and Madras coast, Sri Lanka and Kerala, Konkana, Surat, Sind, Punjab, Kashmir, Afghanistan, Iran, Near East, Turkey, Egypt, Arab Nations, East and South Africa, Ghana, Malaysia, Tibet, Bhutan, Sikkim, Assam and Arakan, Burma, Chinese ports, Pacific Ocean Islands and the Caribbean. In this way, there initiated new trend of civilization in West Europe on 18th century industrial revolution up on worldwide colonialism. New concepts like capitalism, socialism, democracy, rationalism, secular statehood, Marxism, liberalism, individualism, nationalism, Charismatic Personality, Human Rights, class system, equity and global market economy have therefore emerged out along with the technological breakthrough. Civilization always tried to condemn culture and traditional institutions. That directly struck on barter system, exchange of service, caste service, slavery, feudalism-on-estate, joint-extended family system, kinship, various types of solidarity, decision-making authority-on-status, personality, elites, magicians, priests, medicine man, shamans, gender, age-group, village structure, and so forth. Mode of production and reproduction, feed back system in exploitation of nature, notion of biodiversity-and-ecosystem management in traditional way, disaster management as well as dependence on Super-Nature were completely disrupted. The traditional systems from education to health were all blamed inferior. Nature was thought to be regulated by mankind: people have reached to the level of star war, search of oars in other planet, nano technology in robotics, Black Hole and time machine, application of computer in cybernetics and nuclear-new-renewable energy. May there come a day when the human being would extract matter from vacuum in the Space—the Einstein's Ether. So, a direct conflict between folk life and modernity in form of peasantry *versus* urban-industrial development on Global market Economy could lead us in a situation culture opposing civilization. Rajbansis and North Bengal rich in biodiversity could not stay apart from that.

Influence of Peasant Movements

Pabna peasant movement (1872) in northern part of Undivided Bengal against Permanent Settlement System under British Rule as well as earlier agitation made by the Santals of Rajmahal Plains of Jharkhand against the cash-economy then followed by various such revolts in the other tribal pockets of Chota Nagpur, Central India and Andhra parallel to different non-tribal peasant movements were evidenced. All these movements were the substantial proof of the fact that when Social Reformations and Western Civilization were being channeled into the mainstream Indian Society by the middle class intelligentsia plus elites in the urban sectors and the dominant castes in the villages in the 18th-19th century British period; the rural Indian society, especially the tribal stalks, successfully maintained separate existence on culture: they took benefits but only after several modifications and rejection. The western traits could not replace the traditional customs and therefore, the rural sectors have the full power to decide what to accept and which one should be rejected! Secondly, the forest-based tribal communities, agriculturist tribal groups, non-tribal peasants, caste and caste-like groupings, and complex type of rural social stratification in the Indian villages showed their closeness with one another, at least, on the issue of response to civilization input. The British Administration therefore had no option other than to set-up some autonomous pockets for the tribal communities and exclude zones for ancient agricultural groups with some ability of self-determination. They also postulated various protective measures for the notified tribal communities and included them under the Scheduled Tribe Category. The same was done for the lower caste peasant communities who were fallen under the Scheduled Caste category. Later on, in independent India, these provisions were included in the Constitution of India through legislation together with the Human Rights Declaration and other legal provisions.

Post-Colonial Perspective of Indian Society (with special reference to Eastern Part of Indian Sub-Continent including North Bengal)

Still now in Indian society, from the fear of harsh impact of civilization, big-scale development, unidirectional modernity, pure market on cash-credit-equity, rejection of traditional concepts, crisis on cultural identity, mass exploitation of natural habitat and other resources, lack of the traditional alternative as well as and loss of local dialect and language the way of expression of World View; tribal and non-tribal peasant movements have continued to occur in several parts of India; the issue includes the people of North Bengal as part of the state West Bengal and the geographic zone of North-East India:

- **West Bengal:** Movement for protection of land rights, justification in crop sharing, uprooting of the feudalism and farmhouse system, Naxalite movement, Leftist movement, land distribution, three tire village governance system (Panchayat System), decentralization of power, reservation of seats for the lesser gender and the backward categories/Dalits under S.C. and S.T.; empowerment of the poor, protection to the immigrated peoples and minorities, co-operative system, formation of nuclear family system on technological advancement, over-population and increase in additional cost, land-alienation, economic shift from small-scale peasant to tenant and day-labour, labour movement, problem of unemployment, issue of abrupt industrialization, job creation and GDP, lack of Sustainable Development and proper planning, issue of Public-Private partnership, entrepreneurship, nexus of Globalization and anti-Globalization, compiling of all the TKS so to have the Global Public Service of IKS to save the mankind from the negative effect of industry-technological advancement-market economy on nature, need of new kind of Reservation at Global Level, Indigenous Rights, negotiation with the Localized and Isolated people through the advanced sector among them, demand for separate statehood and control over land;

- **North East India:** Demands for separate state, autonomy and self-control on the basis of ethnicity; protest against so-called dominance of mainland India; terrorism and violation of human rights; emergence of regional political parties and separatist outfits; ethnic violence among the autochthons as well as with the invaders; parallel governance; total localization, support at the regional political level; alleged conspiracy against revival of the ancient trade routes helpful for both agriculture and industry.

Rajbansis at the juncture of North Bengal, North Bangladesh (Pabna) and North-east; have developed close attachment to the agricultural plains of North Bengal and other adjoining areas. This could be regarded as the primary step towards considering Rajbansis as Indigenous Peoples of North Bengal with some sorts of land rights in respect of the pre-colonial, colonial and post-colonial period of Indian history.

Level II: Indigenous Knowledge System and Indigenous Peoples

Traditional or Indigenous Knowledge is basically local (*Warren*, 1991). It is actually oral and mostly undocumented (*Ellen* and *Harris*, 1996). It is more practical rather than theoretical and repeating with time. Traditional knowledge traits are generated through informal experiments, intimate understanding of the environment in a given culture and accumulation of generation-wise intellectual reasoning of day-to-day life experiences (*Rajsekharan*, 1993). It is generally tested in the religious laboratory of survival. So, it could be said that Indigenous Knowledge traits are oral, undocumented, simple; dependent over the values, norms and customs of the folk life (>folk culture>material apparatus); production of day-to-day life experience through trial and error; lost and gained as well as asymmetrically distributed. These knowledge traits therefore form a system called the Indigenous Knowledge System (IKS). IKS is quite fragmentarily distributed throughout the globe that is socially clustered. But within a specific society, it looks like the actual knowledge of a given population associated with its World View. There it forms a systematic body of knowledge acquired by the local people. The knowledge bulk seems to be indigenous in respect to a particular geographic area. All of the farmers, landless labourers, women, rural artisans and cattle rearer attached to the region are well informed about their own situations and their resources; what works and doesn't work and how one change impacts other parts of their system. This is really hard to separate the technical part from the non-technical one and the rational domain with the non-rational sector in culture. However, unidirectional input from civilized society at broad scale can create a cultural lag on the traditional way of living. So, whether in favour of Globalization or not, in India there is a need to study applicability and impact of IKS. And to do this, the community-specific TKS has to be thoroughly investigated on humanitarian ground, especially about the agriculture, from the paradox of the complex agrarian rural structure—heterogeneous and extra-caste extra-class extra-power in attitude.

Traditional or Indigenous Knowledge is also regarded by several names, such as, folk knowledge, traditional knowledge, local knowledge, indigenous technical knowledge (ITK), traditional environmental/ecological knowledge (TEK), People's science or ethnology that often look very much confusing and overlapping. IKS is a multidisciplinary subject and incorporates the following dimensions: physical sciences and related technologies, social sciences and humanities (*Atte*, 1989).

IKS as supported by the set of local-level material apparatus from complete lost is also dependent upon non-adaptive domains of the folk life (>folk culture>material apparatus). The folk life is constituted with the altogether assemblage of:

1. non-reflective intangible part of culture (cultural values, social norms, folkways, taboo and traditional belief);

2. reflective and tangible part (set of material apparatus);
3. reflective but non-tangible part (information, knowledge and traditional technologies constituting TKS);
4. mode of communication (formal and informal);
5. the network so formed among agrofacts, artifacts, sociofacts and mentifacts/psychofacts; and
6. the traditional social system (again highly non-adaptive: composed of various institutions).
   6.1. Of the traditional institutions in the social system (non-adaptive), IKS (adaptive) is generally tested in the religious laboratory of survival (religious institution). Overlapping of etic and emic perspectives (from symbols to the World View and again from complex mind structure to ethno-science and folk taxonomy) has to be studied in a qualitative way on post-modern humanitarian ground with deep micro-level study; objectivity should dominate over subjectivity and there would be least amount of biasness during researcher-informant interaction.

So, what has to be taken care off at this level II approach on the quarry of indigenousness of the Rajbansis is the network among IKS of local people, power of words in Rajbansi folk dialect, folk taxonomy, formal and informal communication, institutions, non-functional domains, cultural diversity, modern inputs, local ecosystem and so on. Interrelationship should be worked out among production and technical practices in a specific farming system, conservation of crop-varieties, alternative agricultural production, production of various cash crops/ vegetables/spice/fruit and flower, maintenance of the nutrition level and traditional concepts of health, food preservation, labour-oriented handloom industry, traditional type of division of labour, ethno-fishery, animal husbandry and poultry, agro-forestry and use of forest products (timber and non-timber), sacred grove, agro-ecology and food-web, bio-diversity with feed-back, water and soil management, house construction and kitchen garden, folk-taxonomy, magico-religious performances, belief in Super-Nature, cultural lag, emerging socio-economic challenges, social transformation, sustainable rural and human recourse development. In specific, one has to go through all the domains within the Rajbansi IKS. The major domains are namely, agriculture and post-agricultural practices; animal husbandry and poultry; ethno-fishery; hunting and gathering; artisanship; disease treatment, ethno-medicine and folk remedy; and lastly traditional economic and political system.

All of these domains are now under transformation due to the impact of modernity and unidirectional trends of Global Market Economy.

Even subjected to the civilization input **folk life of Rajbansi community is still composed of whole of the intangible non-adaptive part (culture and norms), the tangible adaptive part (tools, items and goods), thirdly, the intangible adaptive domain (knowledge, information and technology on local basis) and the way of communication.** The latter is again of two major types: informal and formal and be held with people, nature and super-nature via gesture, word, speech, language, art, song, music, dance, drama, play, agricultural and seasonal performances, magico-religious performances/tantra, scripture, chantation/mantra, exchange of message, exchange of goods and exchange of women. For a specific folk life within a definite ecosystem; the reflection of nature on human mind gives rise to symbolic expressions revealing non-adaptive intangible part, tangible and adaptive part and thirdly, intangible part adaptive domain (the last one is very much functional to the non-adaptive domain). The whole process happens according to the psycho-biological functioning of mankind as well as functioning of the social structure organized by various institutions, different social fragments and strata (occupation-wise, gender-wise, status-wise, etc.). Here, the yearly festivals in Rajbansi society symbolizing inner meaning of the folk life and expressing the interrelationship among ecology, energy requirement, traditional technological aid, mode of exploitation of nature, population side and the dependence on super-nature for the Rajbansis are mentioned below:

Mecheni Khela, Tista Buri, Satya Pir, Modon Ka/Bans Khela, Jagannath and Balaram Thakur, Dharam Thakur, Geram Thakur (summer); *Kachibuna, Hudum deo/Benger biao* (monsoon), *Amati, Sotyonarayana, Dodi Kado, Othai pothai, Jitau, Bar a bhasha, Bharar Ghorchhura, Lokhi dak Devi, Boro Debi, Bishahari, Jatra Puja, Bhandani* (spring), *Chondi, Dhap Chandi, Kali Thakur, Khet Uthani (hemanta* or after-spring), *Pushuna, Shial Thakur, Baruni sinan/Maghli sinan, Shiva rati, Gamira Thakur, Chorok, Shiva Chaturdashi* (winter), *Dham, Rakhal Thakur, Shaleshwari, Gorakhnath, Bishua* (autumn); *Jiga Thakur, Tulshi Thakur, Gurbandha Thakur* (anytime in the year) and fear of *Pairy, Jak, Mashan* and other harmful entities. On magico-religious to religious ground, importance has been given up on fish, snake, tortoise, rivers, bird: connectivity is there with the concepts of ancestors' sprit, soul, graveyard, megalith, *Shiva Lingam*, fertility cults, ghosts, angels, malevolent and benevolent entities, deities, black magic, witchcraft, music, musical instruments, song, dance, ornaments, clothing, alcoholism, sexual behaviour, *tantra*, sacrifice, medication, mantra, Sun and fire, day and night, heaven and hell, non-violence and violence, peace and war, defensive and aggressive attitude, hunting and gathering, fishing, domestication and trade, handlooms of wood, bone, bamboo, jute, etc. So, the Rajbansis should have certain level of indigenousness on a multicultural situation despite of all external input. The collection of traditional knowledge traits of the Rajbansis and their systematic functioning would help in bio-diversity management, sustainable development and protection of the ecology.

Level III: Globalization and Indigenous Peoples

In this era of globalization, the rapid developmental activities of the Western-Modern Society have caused 6 major problems as pointed out by UNDP report: Challenges of global warming, Rapid loss of bio-diversity, Crisis-prone financial market, Growing international inequality, Emergence of new-drug resistant disease strains and Genetic engineering (*Kaul, Grunberg and Stern*, 1999). In this regard, Science, Traditional Knowledge and Sustainable Development (ICSU) in the Series on Science for Sustainable Development No. 4 have truly mentioned that the IKS of the local/rural/indigenous communities (basically under the banner of 'underdeveloped' or Indigenous Peoples' category) could be applied so as to control these crises. Loss of biodiversity is one of the four highest risks to natural ecology and human welfare. Biodiversity is a public policy as well as a scientific issue. It is stratified into a four-level hierarchy (i.e., genetic, species, ecosystem and landscape). It maintains ecosystem stability. It provides major cost-free Global Public Services in terms of food, fiber, industrial compounds, fuel and drugs. It has some definite anthropocentric reason so as to preserve different IKS and related cultural diversities. It would ultimately help in the issues of genetic conservation, agriculture and fisheries, forests, wildlife and wetland management (*Cairns and Lackey*, 1992). The highest extinction rate of 1,000-10,000 species per annum since the mass extinction of 65 million years ago has become a serious problem and need help from IKS of the folk people living in nature for thousands of years.

Establishment of small tea gardens at the end of 1990s was highly opposed by the local peasants afraid of land alienation that ultimately led to the initiation of ethnic movement by the Rajbansis in the name of indigenous state Kamtapur. The movement has included the demand of announcing several spoken dialects of the Rajbansis together as separate language *Kamtapuri. A* new branch of separatist movement for the very formation of a new state is thus eventually emerged out. The Kamtapuri Movement is often accused for maintaining nexus with Global Terrorist Network of various Fundamentalist and Extremist groups against the crisis-prone Global Economic System under the control of the West and their allies worldwide. So, the movement might be visualized as an Anti-Globalization process—often going up to the level of Localization and Isolation.

But it might be actually against the Anti-Globalization process of the Leftist Alliance in governance of the state West Bengal since 1977. It might be the scope for a better alternative but under the paradox of Global Market Economy on Capitalism. It might be against the wrong implementation of the large-scale developmental activities as followed by both the present-day capitalists and other alternatives. It would be actually against the issue of severe destruction of biodiversity and exploitation of nature. An appeal in favour of sustainable development could be clearly visualized in this approach. It includes the issue of Globalization in good sense for all the people, especially the worst affected due to severe economic disparity in Global Market System. It speaks out for unification of all the backward people but attached to nature and with belief in super-nature within the common category of Indigenous Peoples, Global Public Service from IKS in favour of two-way Sustainable Development, and thirdly, scope of Global Reservation in the name of Indigenous

Rights. So, communities traditionally against the process of Globalization in this way have been tried to bring in favor of Globalization. This might be one kind of Anti-Localization process through providing certain reservation, development in sustainable way, certain positive roles of Global Market Economy and Globalization of the traditional knowledge traits. The aim is detachment of the local people mainly fallen under the weaker section from the various sub-ordinate politico-economic alternatives of Global Market Economy on Capitalism in this monopolarized world system.

Here the aspect of Indigenous Peoples and Indigenous Rights could be implemented in protection of the land rights provisioned for an indigenous community. But a chance is also there of provocation of a separatist movement on the same issue of land right and participation and share in the development input by the state, the private sector and the civilization machinery. Here output should be Sustainable Development aided by protection of indigenous rights, land rights, basic human rights, betterment of lifestyle, human resource development, poverty eradication, least level of gender discrimination, prevention of biodiversity-loss, protection to the ecosystem, pollution control and feed-back in exploitation of natural resources. What is needed for a proper output is an appropriate negotiation between development input and the World View of local agrarian social structure headed by the Rajbansis of North Bengal (in association with other Bengali castes, local Muslim population, Nepali ethnic groups, mongoloid tribal communities and the Adivasis like Santal, Oraon, Munda, etc.). The basic objectives are here firstly, to understand the Folk Life; to make a bridge between traditional and modern perspectives as well as local and global; to form both formal and informal communication programs; to deal properly with the Multicultural situation, to implement the Reservations and Rules and Regulations according to the Constitution and Law; and lastly, to empower the Co-operatives in service to both Micro-Economy and residues of Traditional Economic System helpful in providing Global Public Service/IKS. A negotiation between emic and etic is highly required in this context.

Rajbansi Social Fold with their IKS could be used in Global Public Service to reduce or replace the harms done by unidirectional and highly exploiting applications of the modern technologies in this current era of Global Market Economy. This ever expanding market system is in search for new customers, more profit, better technologies, more production at low cost, easy access of the energy and raw material, cheep labour and most improved machinery. That could result into the problems like massive unemployment, war for oil and market; struggle for free market economy on equity and credit, capitalist hegemony perishing all other alternatives, pollution and massive exploitation of nature. Information about safe, hygienic, disease-resistant and local varieties within the gene base of local crop-diversity as well as indigenous knowledge traits about every step of cultivation, storage and preservation under the traditional system with feed-back could become very crucial for mankind, climate, biodiversity, bio-geo cycles and food web within an eco-system. So, share of the Rajbansi Agrarian Rural Structure in IKS could play a major role in favour of a harmless or less-harm farming system devoted in protection of the bio-diversity of North Bengal and development input but in sustainable manner.

CONCLUSION

From this above three-level-approach, it is relevant to include the Rajbansi community of North Bengal within the category of Indigenous Peoples but rapidly approaching towards the modernity and at the same time preserving their IKS with the help of non-adaptive domains (symbols). There is a timely requirement to create the scale of indigenousness in respect of the Global Market Economy and the other alternatives the people could have.

REFERENCES

Atte, 1989, In A. Agrawal, 2004, Indigenous and scientific knowledge: some critical comments, *IK Monitor 3(3).*

Babu, S.C., B. Rajsekaran, and D.M. Warren, 1991, Indigenous Natural Resource Management System for Sustainable Agricultural Development, *Journal of International Development* 3(4): 387-402.

Banarjee, S. with D. Basu, D. Biswas, and R. Goswami, 2006, Indigenous Knowledge Dissemination through Farmers' Network: Exploring Farmer-to-Farmer Communication, *In* B. Choudhuri and S. Choudhuri, (eds.), 2007, Indigenous People: Traditional Wisdom and Sustainable Development, IUAES Inter Congress (Vol. 4), New Delhi: IIP.

Brokensha, Slikkerveer and Warren 1993, *Indigenous Knowledge Systems: Cultural Dimensions of Development*, CIESIN.

Ellen, R. and H. Harris, 1996, *Concepts of Indigenous Technical Knowledge in scientific and Developmental Studies Literature: A Critical Assessment*, Electronic Document.

Inge Kaul, Isabelle Grunberg and Marc Stern, 1999, *In* Kelkar, G. with D. Nathan, and P. Walter (eds.), 2004, *Globalization and Indigenous Peoples in Asia- -Changing the Local-Global Interference*, New Delhi. Thousand Oaks, London: Sage Publication.

International Labour Organization, 1991, Convention No. 169 (concerning indigenous and tribal peoples in independent countries, 1989) *In: International Labour Conventions and Recommendations* Vol. 2 (1919-91), Geneva: International Labor Office.

Rajasekaran, B. 1993, *In: Role of Indigenous knowledge in Preserving Biodiversity: Incorporating Indigenous knowledge Systems: A Framework for Incorporating Indigenous Knowledge Systems into Agricultural Research, extension and NGOs for Sustainable Agricultural Development*. CIESEN.

Sanyal, C.C., 2002 (1965), *The Rajbanshi of North Bengal*, Kolkata: Asiatic Society

Warren, D.M., 1991, *Using indigenous Knowledge in Agricultural Development*, Paper No. 127. Washington D.C.: World Bank.

Poverty, Property Rights and Common Property Land Resources Management in the Tribal Belt of Orissa

Anirudha Ojha

In spite of the contributions to the tribal economy and ecosystem, common property land resources (CPLRs) in the tribal belt of Orissa have been degraded and depleted. In pursuance of the management of CPLRs, this study examines efficacy and implications of conferment of property rights for the effective management of CPLRs in the tribal economy of Orissa. The study analyses relative role of the government through plans and policies, incentives and efforts by communities, individuals and voluntary organisations, and ultimately conferment of property rights. Efficient management of CPLR by tribals in their individual capacities and community management need financial support from the state and conferment of property rights without which it would be impossible to reap the desired goal. With a view to managing CPLRs, the government emphasizes nationalization of forests and formulates forest and environmental policies. In spite of this, government failure persists on account of several reasons of which principal ones are abject poverty of the tribals, non-payment of compensation to the tribal households while introducing afforestation on the patch of land otherwise used by them for shifting cultivation and non-assignment of property rights in appropriate form. Centralized management of natural resources including CPLRs by the government has failed to provide right incentives for sustainable resource use. Accordingly, devolution of responsibility and control over natural resources from government agencies to user groups has become a widespread policy trend for their effective and efficient management. On the whole, the present study examines the role and relevance of the existing institutional arrangements for efficient management of CPLR in a tribal pocket of Orissa and has arrived at the inference that all such institutions have failed to meet the desired objective in the absence of conferment of property rights on the one hand and well-defined boundaries of operation on the other. In spite of non-issuance of property rights, the tribal communities are committed to manage CPLRs in order to protect them from further degradation and depletion. Ultimately the study makes some recommendations on the basis of the findings in respect of improving the sustainability of CPLR use in a tribal dominated area.

Key words: Common property land resources, indigenous tribes, CPLR-management institutions, property rights, community-based management, co-management.

I. INTRODUCTION

Poor and their poverty are usually blamed for causing vast degradation of environmental resources especially common property land resources (CPLRs). The most of such blames descends on poverty of the poor in general and indigenous tribes in particular due to their heavy dependence on CPLRs for their survival. Since many social and economic opportunities are not available to forest dwellers, they are compelled to draw their living directly from CPLRs. This act on the part of the poor especially tribes living in and around forests promotes overuse and abuse of these resources leading to their degradation and depletion. The incidence of heavy dependence results in overexploitation and degradation of CPLRs particularly forest and increasing hardship to the poor who depend heavily on these resources leading to the intensification of environmental degradation. When property rights on natural resources are absent and unenforced, no individual stakeholder bears the full cost of resource degradation resulting thereby overexploitation and tragedy of the commons. The 'tragedy of the commons' has been used to explain overexploitation of forests and fisheries, overgrazing, air and water pollution, abuse of public lands, population problem, extinction of species and other problems of resource misallocation. A circular causation exists between poverty and environmental degradation. With high discount rate, short planning horizon and inadequate property and assets of their own, poor people invest less or no capital for proper up-keeping of natural resources and averting their degradation. The hypothesis of 'poverty of people is a cause of environmental degradation' has not always been maintained. The environmental degradation affects the poor more than the rich as the former has no adequate sustainable means of livelihood to protect them from environmental resource degradation. They have been affected severely by the natural shocks like floods, droughts and cyclones.

Common property land resources (CPLRs) have multiple competing uses for the livelihood sustainability of indigenous people living in poverty and deprivation in the ecologically fragile regions. They are the sources of contributing and providing food, fodder, fuel, and materials for house construction, medicinal herbs and other timber and non-timber forest products to the household economies. A significant proportion of household income of the poor and employment generation is derived from CPLR-based activities particularly forest CPLRs. In spite of such contributions to ecosystem of the indigenous people on the one hand, and protective measures and policies pursued by the government on the other, tragedy of the commons has been experienced in the Scheduled Areas in Orissa, which threatens survival of tribals depending on common forests. Among the likely primary production activities, most of the tribals (Juang and Pauri

Bhuinya) in the study blocks of Keonjhar district in Orissa practise shifting cultivation (podu or swidden) for their survival. On account of non-availability of timely employment opportunities, swidden is practised by indigenous tribes till recently as a culture and traditional practice and occupation without considering its adverse effects on forest ecosystem and their economy. Without this practice their economic and social livelihood would go on to deteriorate. Poverty of people, absence of well-defined property rights, population growth, commercialization of CPLR products, practice of shifting cultivation, and depletion of social norms and tribal traditions are the factors responsible for causing adverse environmental impacts in tribal areas and pockets. These factors are due to institutional failure, lack of clarity or enforcement of rights of access to and use of resources, policies which provide disincentives for resource conservation, market failures, dissonance of local level institutions and governance constraints. The problem which arises out of the poverty-environment nexus is the problem of management of natural forest resources which will have to be protected from overuse and degradation and at the same time by allowing the indigenous people to go with their own cultural patterns, social institutions and legal systems. Among the likely strategies or policy options or institutions that set rules to govern the use of CPLRs by a particular group of users (to the exclusion of others) for their management, specification of exclusive property rights is an important and effective means of managing CPLRs in the economy. With well-defined property rights and well-demarcated resource boundary, individuals, communities, government and NGOs should embark on necessary efforts and incentives for the internalization of negative externalities and effective management of CPLRs without hindering livelihood sustainability of the indigenous people.

Following are the major objectives of the current work:

1. To examine lack of assignment of secure property rights on CPLRs, despite poverty of tribals and inadequate infrastructure, which is a cause of depletion and degradation of CPLRs and hence the environment.
2. To study the likely implications of the conferment of secured property rights on CPLRs particularly to the indigenous tribes.

3. To examine the efficacy of well-defined property rights on CPLRs management, especially the management of forest CPLRs.

In order to achieve these objectives, field survey data used. Following are some of the working hypotheses of the current research:

1. That, poverty of the tribal people, and inadequate private property rights and assets cause heavy dependence of the tribals on common forests for collection of their livelihood needs.
2. That, in the management of common property land resources in the tribal economy there is complete failure of government and its policies, planning and the market.
3. That, between poverty and non-issuance of adequate property rights to tribals, latter is a greater cause of depletion and degradation in the quantity and quality of CPLRs.

A set of economic, demographic, social, infrastructural and environmental parameters have been chosen to select two scheduled areas of Keonjhar district from among the existing scheduled areas. On this basis, two Community Development Blocks namely Telkoi and Banspal which are inhabited mostly by two indigenous tribes Juang and Pauri Bhuinya constitute the study areas. Moreover, largest area of each of these two blocks is under forest cover. These two indigenous tribes depend directly on CPLRs for their social, economic and cultural survival. A four stage random sampling procedure has been used to reach the ultimate sample units. These sample units have been solicited their responses to several questions contained in an open-ended structured questionnaire covering the aspects of the study; of course, the open-endedness has been limited to a few questions. Though the study depends primarily on the data collected through questionnaires from these two sample areas, the secondary data collected from published sources and compiled from the records of several organisations and Departments of Government of Orissa and India are used for historical sketch of the study area. So far the analysis of data is concerned, in most of the cases, tables, percentages, proportions and averages are used.

II. Study Results and DISCUSSION

II.1 Tragedy of the Common Property Land Resources in the Tribal Economy

'The tragedy of the commons' has been experienced in the two scheduled blocks of Keonjhar district due to: (i) decline in CPLRs area and number of products obtained by indigenous tribes, (ii) nearly total extinction of rich vegetation species from a number of CPLR-compartments, (iii) depletion in grazing potentials of CPLR, (iv) diminution in number of watering points, and (v) perceptible variations in density of trees and shrubs on protected and unprotected CPLR-compartments. A National Sample Survey Organisation Report (1999) estimates that in Orissa CPLRs form 11 per cent of the total geographical area of the state, rank seventh among all states in India. Table 1 presents a detailed

TABLE 1

Availability of the Area Under CPLRs and Forests in Keonjhar District

Sl. No.	*Particulars*	*Area*
	Total Geographical Area (000 ha)	830
	Total Forest Area (000 ha)	310
I.	Total CPLRs (000 ha.)	251
	Total Forest CPLRs (000 ha.)	126
	Total Non-forest CPLRs (000 ha.)	125
	Total Forest Area as Percentage of District Total	37.30
	CPLRs as Percentage of Total Geographical Area	30.24
II.	Forest CPLRs as Percentage of Total Geographical Area	15.18
	Non-forest CPLRs as Percentage of Total Geographical Area	15.06
	Per capita CPLRs in the district (ha)	0.16
	Per capita CPLR with respect to tribal population (ha)	0.36
	Per capita non-forest CPLR w.r.t. tribal population (ha)	0.18
III.	Per capita forest-CPLR w.r.t. tribal population (ha)	0.18
	Per capita forest area w.r.t. tribal population (ha)	0.45
	Per capita land area w.r.t. tribal population (ha)	1.20

Source: Directorate of Agriculture, Bhubaneswar, and PCCF, Bhubaneswar.

TABLE 2

Area of Lands Used as Commons in Keonjhar District

(*in 000 ha.*)

Sl. No.	*Land Use*	*1975-76*	*1980-81*	*1985-86*	*1990-91*	*1996-97*	*2000-01*	*2005-06*
1.	Total Geographical Area	830	830	830	830	830	830	830
2.	Forests	232 (27.95)	300 (36.15)	250 (30.12)	250 (30.12)	310 (37.35)	310 (37.35)	310 (37.35)
3.	Forest CPLR	52 (6.27)	118 (14.22)	66 (7.95)	78 (9.40)	50 (6.02) {15.31}	50 (6.02) {15.29}	50 (6.02) {15.29}
4.	Non-forest CPLR	71 (8.55)	71 (8.55)	157 (18.92)	124 (14.92)	125 (15.06)	112 (13.49)	157 (18.92)
	Total CPLRs	123 (14.82)	189 (22.77)	223 (26.87)	202 (24.32)	175 (21.08) {30.36}	162 (19.52) {28.80}	207 (24.94) {34.21}

Figures in bracket ( ) show percentage to total geographical area of the district.
Figures in bracket { } show percentage to total geographical area of the district which includes other forests under Revenue Department.
Source: Directorate of Agriculture, Bhubaneswar, and PCCF, Bhubaneswar.

estimation on availability and size of CPLRs in Keonjhar district. Excluding all government reserved forests and revenue land, CPLRs in the Keonjhar district form 30.24 per cent of its total geographical area. The per capita CPLRs in Keonjhar district is 0.16 hectares, while per capita CPLRs available to a tribal household in the district work out to 0.36 hectares. Perusal of Table 2 shows the change of CPLR, forest area, forest CPLR and non-forest CPLR over time. It is shown that there is considerable decline in area of the forest as percentage of total geographical area of the district. And the area of non-forest CPLR as percentage of geographical area of the state has gradually increased over time, showing the sign of large scale encroachment, shifting cultivation, deforestation and environmental concerns. Despite such variation in the size of CPLRs over time, the present field survey noted significant depletion and decrease in the area of community pasture and grazing grounds, village forests and woodlots and village site and the area of dense forest on which they depend (Table 3). Perusal of Table 4 noted the visible degradation of CPLRs in the study blocks. A drastic decline in the number of products has been reported following the disappearance of a number of plant and tree species in tribal areas which the tribals used to gather from the commons. Animal species including mostly tiger, bear, elephant and the like have decreased in number. An important form of visible degradation involved in terms of gradual increase in distance between the land of habitation and

TABLE 3

Forms of CPLRs Degradation in the Tribal Economy

Sl. No.	*Form of Degradation*	*Households Reporting (in per cent)*	
		Banspal	*Telkoi*
1.	Reduction in CPLR area	100	100
2.	Decrease in number of trees and plants	100	100
3.	Depletion in grazing potentials	100	100
4.	Decrease in number of watering points (ponds/tanks)	36	68
5.	Decrease in number of CPLRs products	100	100
6.	Decrease in animal species	100	100

*Data in the Table has been collected from the Primary Source.

TABLE 4

Visible Degradation of CPLRs in the Tribal Economy

Sl. No.	*Visible Degradation*	*Banspal (% of Tribal Households)*	*Telkoi (% of Tribal Households)*
1.	Disappearance of Species (Animals, Trees, and Plants)	100	100
2.	Increase in Distance of CPLRs from Land of Habitation	100	100
3.	Average Increase in Distance (in kms)	3.77 (377)	4.25 (425)
4.	More Time Required for Collection of CPLRs Products	100	100
5.	Change of Authority Over Use of CPLRs and Their Products	99	100
6.	Causes for Change of Authority:		
	(a) Take Over of Forest Department	96	100
	(b) Village Level Conflict	14	5
	(c) VFC Participation	Nil	12
	(d) VSS Participation	Nil	15
	(e) Commercialisation of CPLRs for Sericulture	1	4
7.	CPLRs Use in Different Festivals and Occasions	100	100
8.	Members Using CPLRs in Different Festivals and Occasions:		
	(a) Alone	Nil	Nil
	(b) With Village Members	100	100

Data in the Table has been collected from the Primary Source.

CPLRs. This situation is perpetuated in terms of average increase in distance of 3.77 kms in Banspal and 4.25 kms in Telkoi (Table 5). Among the factors causing such tragedy, the prominent ones include poverty and heavy dependence of tribals, underdeveloped infrastructure, non-specification of secured property rights, large scale encroachment for shifting cultivation and habitation, population growth, commercialisation of CPLRs products, and government failure and market failure (Table 6).

TABLE 5

Distance Travelled by Tribal Households for Collection and Sale in the Market

Sl. No.	*Distance Between*	*Average Distance Covered by Tribal Households (in kms)*		*Average Time Spent by Tribal Households (in hours per day)*	
		Banspal	*Telkoi*	*Banspal*	*Telkoi*
I.	CPLRs and Place of Habitation	3.84	4.50	1.13	2.07
II.	Market and Place of Habitation	15.42	15.70	2.59	2.24

Data in the Table has been collected from the Primary Source.

TABLE 6

Causes of Degradation of CPLRs in the Tribal Economy

Sl. No.	*Number of Households Reporting Causes of Degradation (in per cent)*	*Banspal*	*Telkoi*
(i)	Increase in number of livestocks	24	66
(ii)	Increase in number of human population depending on CPLRs	100	100
(iii)	Felling of trees	100	100
(iv)	Increase in marketing of CPLRs products	95	100
(v)	Illegal Privatisation (Encroachment) of CPLR	100	100
(vi)	Shifting Cultivation	98	38
(vii)	Increase in Agricultural Development (Mechanisation of Agriculture)	46	76
(viii)	Failure of village level authority system	69	94
(ix)	Depletion of social capital (social trust, norms and network)	100	100
(x)	Allocation of CPLRs for development of industries	21	02
(xi)	Absence of property rights over CPLRs	100	100

Data in the Table has been collected from the Primary Source.

II.2 Poverty of Tribals, Property Rights and CPLRs Degradation

It is evidenced from the two sample spaces that poverty of the tribals with inadequate endowments and property exert considerable influence on dependence of tribals on CPLRs for their livelihood needs. In the two sample spaces, tribals having high rate of time preference and shorter planning horizons which prevent them from investing capital for the preservation of CPLRs leading to their depletion and degradation. Even if the percentage of average annual income from CPLRs is higher, they derive less from CPLRs in absolute terms as compared to households having private owned land of more than 4 acres. Household with higher privately owned land, adequate endowments and livestock assets derive a greater share from the common forest lands. Most of the tribals who are marginal cultivators possess privately owned land of 1 to 3 acres in both the study blocks encroach upon vast areas of forest land for habitation and cultivation and practise shifting agriculture which is environmentally destructive. Despite greater reliance on CPLRs for livelihood needs and have less or negligible investment incentives for preservation of CPLRs, the encroachment of vast areas of common forest lands for shifting cultivation and habitation leads to their depletion and degradation. Due to socio-economic backwardness of the tribals, inadequate irrigation facilities and non-reaching of advanced technology, agriculture in the two sample blocks is highly dependent on traditional technology like wooden plough, hoe and axe. A high degree of linkage between CPLRs and tribal agriculture of both settled and shifting on hill slopes and terrains has been experienced. Collection of tree and grass fodder and livestock grazing are widespread and uncontrolled in the two sample spaces. Poverty-induced population growth among tribals not only increases encroachment of forest lands for cultivation but tribal households are also engaged in gathering and collecting CPLRs products for self-consumption and sale.

Like poverty of tribals, incomplete, unenforced or inconsistent property rights on the one hand and absence of well-demarcated boundaries on the other are mainly responsible for tragedy of CPLRs. After the take over of the forest department, forests of all categories including protected, unclassed and reserved category are declared as state property over which the government claimed its ownership rights for their management. The customary tenure systems of using and managing the common forests in the tribal economy have been neglected. Long standing customary rights and occupancy of forests by the tribals have been ignored by the forest department with the implementation of forest and environmental policies and Acts of protecting forest species and wildlife and at the same time provides protection of these resources from degradation. However, the tribals have not been displaced from their land of habitation rather they dwell around and inside forests and extract products from CPLRs for self-consumption and sale beyond their carrying capacity. Despite of their customary tenure systems, there is no issuance of property rights by the revenue department, as most of common lands in the tribal areas are under forest department. Issuance of property rights creates conflicts between the tribals and government which is neither resolved nor colluded till date. Even if the tribals applied for issuance of permanent property rights over encroached upon common forest lands and agitated against the government for resolving their long standing demand, the revenue department fails to issue rights to tribals on such lands as these lands are under forest department. This creates uncertainties on the part of both the government and the tribal communities in the two sample spaces towards sustainable use of CPLRs. The consequence of non-issuance of secure property rights, conflicts between the government and the tribals, and failure of the customary tenure systems in the tribal economy have been reported as a stronger cause of CPLRs degradation than the poverty of the tribals.

II.3 Rights and Duties of Tribals and Preservation of CPLRs

Living around and inside the forests since time immemorial the tribals claimed their customary and usufruct rights over forests for habitation, cultivation and collection of products from forests for their socio-economic and cultural survival. And at the same time community level institutions of tribals manage CPLRs efficiently through their traditions, social habits and culture. They attached sacredness to certain tree, plant and animal species for preservation, viz., species having medicinal values, closely related to their life, fruit bearing species, not cutting trees and plants in different festive occasions, species having religious values, young plants and species, and the like. The constructive dependence of the tribals in the past through their traditions, culture and habits had enough ability to regenerate and manage CPLRs and species without going beyond the carrying capacity of the environmental resources. Accordingly, they had exclusive rights over the use of CPLRs in the absence of well-demarcated boundaries and other *a priori* conditions of managing CPLRs excluding successively the non-owners. There was strong cooperation, coordination, leadership and social norms and trusts within the framework of communal authority systems of using and managing CPLRs in the tribal economy which placed adequate restriction, restraint and punishment for the violators within and even outside the group. Over time, communal ownership and usufruct rights of the tribals have been restricted and curtailed and even delegitimised due to penetration of colonial administration, traders, money lenders, market forces, developed transport and communication, establishment of development and mining projects, reservation of forests as protected and reserved category. And also rapid population growth in tribal areas, large scale encroachment of forest lands for habitation and shifting cultivation, and environmental legislations on forests and wildlife have been a threat to customary rights of the tribals and sustainable use of CPLRs. It is unfortunate to note that the mutual constructive dependence of tribals through their traditions, customs, myths and beliefs has been changed or eroded over time. In particular, forest policies during the colonial administration restricted the rights of the tribals in the name of commercializing the forests and forest products and the same continued till the enactment of post-independence NFP 1952. In spite of the constitutional provisions, tribal sub-plan allocations, and concessions issued in different forest policies, there is failure of the government, market and policies to restore and preserve customary rights of the tribals over CPLRs. As such, there is conflict of interest between the tribals and government for specification of secure and exclusive property rights as the former can not be displaced from their land of habitation and subsistence means of livelihood. With gradual integration of the tribal economy with the mainstream economy, society and culture, they have been exploited, neglected, displaced and dispossessed from their perennial source of livelihood.

Consequent upon the above-mentioned problems that influence rights, duties and livelihood of tribals, the recent NFP 1988, PESA Act, 1996 and Tribal Rights and Recognition Bill, 2006 have recognized rights of the tribals and emphasized participatory management of environmental resources.

II.4 Property Rights and Management of CPLRs

One of the *a priori* conditions for CPLRs management is the specification of well-defined property rights. Common property access having no well-defined property rights is subject to emergence of negative externalities and non-optimal market allocation, i.e., the market failure. In the present study, the policy prescription of creating and enforcing property rights fails to achieve desired goals of arresting CPLRs from degradation. As mentioned earlier, poverty of the tribals and inadequate property and assets, and inaccessible infrastructure facilities in the fragile tribal ecosystem exert considerable influence on overuse of CPLRs and hence the environment. Individual level of rationality in appropriating CPLRs in the tribal areas beyond carrying capacity of such resources lead to collective irrationality and non-cooperation and hence the tragedy of commons. The market-based instruments of internalising negative externalities arose out of the intricate connection between poverty of tribals and their dependence on CPLRs fails to achieve social optimum of resource allocation and management. The state government fails to impose taxes on the generator of externality as most of the tribal households are deprived of private assets, infrastructure and living in fragile environment.

In the two sample spaces, tribals have extensive rights of access not only on pasture and grazing lands, and village forests which are under the jurisdiction of the village or village panchayat but also on forest-CPLR under forest and revenue departments. Apart from these, there is a system of customary user and ownership rights over CPLRs which shape their tradition, social habits and culture. As the British entered into the territory, restrictions were placed on tribal rights over free use of CPLRs. Even after independence the Government of India adopted the policies of forest use and management as envisaged by the British. Consequent upon the take over of forests by the forest department and their regulation through forest policies and legislations,

TABLE 7

Property Rights and CPLRs

Sl. No.	Property Rights and CFLRs Management	Banspal (% of Households)	Telkoi (% of Households)
I.	Tribal Households Reporting Management of CPLRs	95	98
	Mode of Management of CPLRs in the Tribal Economy:		
	(i) Management by Community such as tribal communities, VSS and VFC	90	96
	(ii) Management by Individual Households	15	04
II.	(iii) Management by the Competent Authority such as Forest Department/ Revenue Department	32	83
	(iv) Management by Voluntary Organisations/NGOs	11	20

Data in the Table has been collected from the Primary Source.

TABLE 8

Problems Faced by the Tribals in Community Management in the Absence of Secured Property Rights

Sl. No.	Particulars	Banspal (% of Households)	Telkoi (% of Households)
I.	Need Financial Help from the Government for Guarding and Watering	62	74
II.	Absence of Proper Cooperation and Coordination between Forest Department and the Community	24	26
III.	Problem of Large Scale Encroachment	17	20
IV.	Large Scale Forest Fire by the Tribals for Shifting Cultivation	20	16
V .	Theft of Cutting Trees at Night by the Outsiders	27	6
VI.	Intra-Village Level Conflicts	06	10
VII.	Absence of Cooperation, Unity, Coordination and Leadership among the Community Members	04	05
VIII.	Greater Influence of Depletion of Social Capital	02	05
IX.	Extreme Poverty and Inadequate Assets and Land Holdings	05	03
X.	Commitment to Meet Basic Needs	20	11

Data in the Table has been collected from the Primary Source.

several restrictions have been placed on forest-CPLR use despite of certain concessions. There is no evidence of specification of secure and exclusive property rights to the tribal households over CPLR compartments. Table 9 neatly exhibited specification and assignment of property rights in the management of CPLRs. In spite of their long standing occupancy of encroached upon forest lands for habitation and shifting cultivation, there is complete failure of the government to assign secure and exclusive property rights. Failure to assign property rights by the government often leads to conflict between the tribals and Forest Department and Revenue Department which has not yet been settled or colluded. Insecurity of ownership, mismatches between the state and indigenous forms of ownership, and unequal distribution of

TABLE 9

Conferment of Property Rights and CPLRs Management

Sl. No.	*Particulars*	*Banspal*	*Telkoi*
I.	Conferment of Exclusive and Secure Property Rights by the Competent Authority or Existence of Secured Property Rights	Nil	Nil
II.	Households applying for conferment of property rights (in per cent)	51	51
III.	Consequence of such applications:		
	(i) Rights Issued	-	-
	(ii) Rights Not issued	51	51
IV.	Household level conflict with the forest department and revenue department		
V.	Outcome of Conflict:		
	(i) Complete withdrawal	-	-
	(ii) Collusion	-	-
	(iii) Not settled	16	10
VI.	Cause of Possessing and Managing CPLRs by the Tribals in the Absence of Conferment of Property Rights by the Competent Authority:		
	(i) To meet livelihood needs of present generation	05	12
	(ii) To meet livelihood needs of future generation	69	75
	(iii) To meet livelihood needs of both the generation	22	20
	(iv) To save CPLRs from degradation further and extinction of species and biodiversity	05	12

Data in the Table has been collected from the Primary Source.

ownership and use rights are frequent sources of conflicts and poor envirónmental decisions of managing CPLRs in the tribal areas. In spite of non-specification of property rights by the government and abject poverty of the tribals, the households of tribals (about 95% in Banspal and 98% in Telkoi) however reported the presence of local level management institutions including tribal communities, individual level efforts by the tribals, the government and voluntary organizations for the management of CPLRs (Table 10). Perusal of Table 10 reveals the active participation of communities like the tribal communitarian institutions, Vana Samrakshana Samiti, Village Panchayat and Village Forest Protection Committee. This is confirmed by more than 90 per cent of the tribal households in both the sample spaces. As far as management of CPLRs by the community is concerned, several problems have been encountered in the two sample spaces (Table 8). Table 8 noted such constraints as poverty of tribals, need for financial assistance, inadequate cooperation and coordination between the communities and the forest department, large scale encroachment and forest fire, absence of well-defined property rights and well-demarcated boundary that are conducive to community management. Since poor tribals derive greater share of income from CPLRs, in the event of tragedy of CPLRs they bear greater hardship. The tribals face greater risk from environmental degradation as they live on marginal and fragile lands prone to

TABLE 10

Institutional Arrangements for the CPLR Management

Sl. No.	*Institutions Managing CPLRs*	*Banspal (in per cent)*	*Telkoi (in per cent)*
I.	Households Reporting Presence of Institutional Arrangements such as:	94	98
II.	Type of Institutions		
	(1) Village Panchayat	05	04
	(2) Village Forest Protection Committee (VFC)	.79	77
	(3) Vana Samrakshana Samiti (VSS)	14	22
	(4) Forest Department (FD)	34	92
	(5) Individual Stakeholders	12	15
	(6) Voluntary Organisations like NGOs	25	28

Data in the Table has been collected from the Primary Source.

natural shocks of drought, land degradation and erosion. Despite of the factors affecting adversely the functioning of community management, the tribals felt insecure due to non-specification of exclusive property rights by the government (Table 9). There is conflict of interest between the government and the tribals for conferment of property rights. It is unfortunate to mention that the outcome of such conflicts has not yet been settled nor government issued exclusive property rights over encroached upon forest lands for habitation and shifting cultivation. In spite of non-issuance of property rights, the tribal communities are committed to manage CPLRs in order to protect them from further degradation and depletion. In the two sample blocks, about 70 per cent of households confirmed that they manage CPLRs to meet the needs of future generation signifying sustainability of environmental resources.

In addressing 'the tragedy of the commons' in the tribal economy, it is observed that policy measures must not only focus on specification of property rights, but also carefully consider the context in which property rights regimes are to be placed and the extent to which they are enforceable. The need for assignment of property rights in the two sample spaces is due to overuse, depletion and degradation of CPLRs arising from misleading price signals which result from the absence of markets in resource and environmental assets. With exclusive and secure property rights, resource depletion is internal to the owners or users, while under open access it is external to the users. Without clear property rights, environmental resources cannot be sustained over time, but property rights alone cannot ensure use levels that reflect use and protection of resources. In the continuum of property rights, the government claimed state rights over forests after take over of the forests and classifying forests as protected, reserved, unclassed and forests under revenue department. Accordingly the ownership lies with the state for which the government protects and regulates use and ownership over forests. At the same time, the tribals in the two sample spaces claimed their community control over such forests as they are the original inhabitants of the land. Therefore, both the government and community of tribals influence ownership and use rights over forests to manage and regulate, and ultimately former being the supreme authority has greater influence on the management of CPLRs.

II.5 Property Rights and Institutional Arrangements of Managing CPLRs

Among the prominent existing institutions mention has been made of individuals, communities, government, voluntary organisations and market. With a view to managing CPLRs, the government emphasizes nationalization of forests and formulates forest and environmental policies. Plan-allocations to tribal sub-plan, financial assistance to tribal development agencies, constitutional provisions and poverty alleviation measures are operational in the scheduled areas of Keonjhar district to develop socio-economic life of the tribals and infrastructure in tribal areas, preservation of their culture, tradition and habits and protection of the environment. In spite of the government provisions, policies and environmental legislations, the government fails to arrest severe degradation and depletion of CPLRs and the environment due to abject poverty of the tribals, large scale encroachment, forest fire and shifting agriculture. Low socio-economic and HDI indicators, marginal and fragile living conditions in the forest, inadequate private endowments and high discount rate among tribals in the tribal areas tend to violate the government policy measures of protecting the environment. Non-payment of appropriate compensation and rehabilitation measures while introducing afforestation programmes on lands otherwise used by the tribals for shifting cultivation fails to achieve goals of controlling large scale forest fire and shifting cultivation.

In response to failure of centralized management by the government and of the market in the absence of conferment of secure property rights and well-demarcated boundary, the management regime has been shifted to community-based management and comanagement. Community-based management institutions that are involved in managing CPLRs in the tribal economy include Village Panchayat, norm guided and tradition bounded institutions, Village Forest Protection Committee (VFPC) and Vana Samrakshana Samiti (VSS). For the efficient management and participation in community management, the community members need financial support from the state, conferment of property rights, well-defined boundary to prevent intra-village level conflict and theft of cutting trees, and coordination and cooperation between individuals, communities and the government in appropriate form, without which it would be impossible to reap the desired goals of managing CPLRs by the community. The current study reveals that about 34 per cent of households in Banspal and 67 per cent in Telkoi reported the effectiveness of management of the CPLRs by the community. Like government failure, there is failure of the community to manage CPLRs resulting from inadequate financial assistance, non-assignment of property rights, non-cooperation among community members, absence of well-defined boundary, depletion in social capital, abject poverty of the people, large scale encroachment and shifting agriculture. As such participatory management through people's participation and collective action is the most delicate one to deal with 'the tragedy of the commons'. Accordingly, many scholars advocated that decentralized collective management of CPLRs by their co-users is the most appropriate natural resource management (Olson, 1965, *Runge*, 1986, *Wade*, 1987, *Chopra et al.*, 1990, *Ostrom*, 1990, *Kadekodi*, 1991, *Lise*, 2000, *Adhikari*, 2005). An important participatory approach of managing forest at the grass-roots level that exists in India since 1970s is the joint forest management (JFM). In August 1988, Orissa became the first state to involve local people officially in the protection of natural forests. Under the JFM scheme, the tribal community in the study blocks of Keonjhar district participated with VFPCs, VSS, Village Panchayat and the Forest Department taking decisions of managing the forest CPLRs and most often mediated by NGOs. Tables 12 and 13 exhibit the existence of JFM and extent of CPLR management in the two sample blocks having implications on livelihood sustenance of the tribals. Under the regeneration programme, cooperation and coordination are reported to have been nil in most of the tribal villages due to: (i) CPLRs are managed exclusively by the tribal communities, (ii) objectives and functions of JFM and local level institutions are different, (iii) no control over podu cultivation, (iv) poverty and deprivation of tribals, (v) indigenous tribes made timber felling and selling of CPLR products as their occupation, (vi) conflicts within the forest department due to lack of proper training and reduced frequency of visits to JFM area. Conflicts within the local community institutions arise due to inadequate representation of all sub-groups and interests, inequitable sharing of costs and benefits from managed-CPLRs under JFM among the sub-groups, theft of cutting trees at night and inadequate finance by the forest department.

From the ongoing discussion of community and participatory management, it is noted that both the management regimes have taken active incentives of managing CPLRs through attending in the decision-making, planting tree species, fencing and guarding of CPLRs plots (both regenerated and degraded), prohibiting outsiders not to cut trees for sale, and restriction on the practice of shifting cultivation. An important element of strong financial assistance and vigilance by the forest department has found no mention in the community management. A systematic rotational use of forest products by the users on weekly basis and complete prohibition of felling of trees for sale is an added advantage in comanagement. On the other hand, in community management of CPLRs, the local communities are norm bounded to maintain unity and cooperation among the community members. Finance is an important constraint in community management. However, community management in the two sample spaces is more comfortable with regard to use and sharing of products, and restoration of tribal rights as most of the tribals are dissatisfied with the government for the non-specification of property rights. Sometimes there is conflict of interest between the community and the government with regard to issuance of property rights on their encroached land for habitation and cultivation. The very secret of favouring the community level management by the tribals in the two sample spaces is not to disclose their entire encroached land and sometimes there are fear of withdrawal by the government and likely displacement due to afforestation programme. However, both the management regimes require more financial assistance from the government to meet shortage of irrigation, technology and fertilizer used. It is observed from Table 11 that about 56 per cent of households in Banspal and 34 per cent in Telkoi preferred community management as an effective management institution of managing CPLRs. One of the important factors causing the tribals not preferring comanagement is the conflict that arose between the community and the government for non-issuance of property rights. And also they blamed the government because of inefficiency in protecting the locally situated CPLRs without paying compensation measures in

TABLE 11

Relative Advantage and Disadvantage of Preferring Community Management and Co-Management by the Tribals

Sl. No.	*Particulars*	*Banspal (% of Households)*	*Telkoi (% of Households)*
I.	Household Preferring Community Management to Co-management	56	34
II.	Causes of Preferring Community Management by the Households:		
	(a) Conflict with the Government for conferment of property rights	16	14
	(b) Better unity is maintained among community members	50	32
	(c) Local community is better aware of village commons	50	34
	(d) Inefficiency of FD in Protecting CPLRs	31	33
	(e) Taking better afforestation programmes by planting trees	28	32
III.	Household Preferring Co-management to Community Management:	28	61
IV.	Causes of Preferring Co-management by the Households		
	(a) Providing Financial Support	26	56
	(b) Deepening unity among community members	19	13
	(c) FD/Govt. is more efficient than any other community	25	50
	(d) Inadequate vigilance by the community	19	46
	(e) To control forest fire	5	23
	(f) To provide/solve water problem	5	54
	(g) Building confidence, unity and integrity among users for better management	12	20

Data in the Table has been collected from the Primary Source.

appropriate form. About 50 per cent of households in Banspal and more than 32 per cent in Telkoi reported that CPLRs is managed due to unity among the community members, and members are well-acquainted with problem of the local commons and protection thereof. Some of the tribals are satisfied with the afforestation programme undertaken by the community. As contrast to community management of CPLRs, about 30 per cent

TABLE 12

Participatory Management of CPLRs in Banspal

Study Blocks	*Gram Panchayats*	*Villages/ Hamlets*	*Existence of JFM*	*Type of Local Level Institutions Ensuring JFM with FD*	*Extent of CPLRs Management*
Banspal	Jatra	Badakul	No	Nil	Nil
	Raigod	Podadihi	Yes	VFC	Strong (over podu)
	Suakati	Danla	Yes	VFC	Limited
	Talachampei	Talachampei	Yes	VSS	Strong
		Thakurdihi	Yes	VFC	Limited
		Rugudi	Yes	VFC	Limited
		Kantakoli	Yes	VFC	Limited
	Bayakumutia	Bayakumutia	Yes	Panchayat VFC VFC	Limited
	Karangadihi	Tangarpada	Yes	VFC	Limited (no control over podu)
	Gonasika	Gonasika	Yes	VSS	Strong (but no control over podu)

Data in the Table has been collected from the Primary Source.

of tribal households in Banspal and more than 60 per cent in Telkoi have favoured comanagement of managing CPLRs. And 25 per cent in Banspal and 50 per cent in Telkoi supported that forest department or the government is more efficient in managing CPLRs than any other management institution. This is due to strong financial support by the government, vigilance by the government officials having adequate knowledge of dealing with the tragedy, and has the ability to control large scale forest fires and podu, and building confidence, unity and integrity among the user groups. Many tribals have not supported community management due to depletion of social capital, decrease in welfare activities in the villages, greater influence of political affiliations and village level conflicts. It is therefore concluded that community-driven comanagement is the delicate one to deal with

TABLE 13

Participatory Management of CPLRs in Telkoi

Study Blocks	*Gram Panchayats*	*Villages/ Hamlets*	*Existence of JFM*	*Type of Local Level Institutions Ensuring JFM with FD*	*Extent of CPLRs Management*
Telkoi	Saleikena	Jansanpur	Yes	VFC	Limited
		Saleikena	Yes	Panchayat, VFC	Limited
	Jagamohanpur	Jagamohanpur	Yes	VFC	Limited
		Sinkulabahal	Yes	VFC	
		Saruali	Yes	VFC	
	Oriya	Oriya	Yes	VSS	Very Strong
		Sapnanji	Yes	Nil	—
		Katarapali	Yes	Nil	—
	Khuntapada	Pitanali	Yes	VFC	Limited
		Kuladera	Yes	VSS	Limited

Data in the Table has been collected from the Primary Source.

the tragedy of the commons and greater environmental concerns.

Consequent upon the relative advantages and disadvantages of centralized management, community management and comanagement, and uncertainty that arose out of non-issuance of secured property rights, the tribals at their individual capacity level have taken a series of measures for the management of local commons. Among the notable incentives taken by tribal households, mention has been made of planting of trees species on their own land and encroached common lands, preventing livestock from grazing, preventing non-owners and outsiders to cut trees and harm species, fencing and guarding of the commons and influencing morality of others. However, efficient management of CPLRs by tribals at their individual capacity level needs financial support from the state, conferment of property rights without which it would be impossible to reap the desired goal, well-demarcated boundary to prevent the outsiders from cutting trees at night, and control of large scale encroachment and forest fire for shifting cultivation. Consistent with 'the tragedy of commons' in the two sample spaces and failure of the government and market, non-governmental organisations (NGOs) have played an important role in the socio-economic development of tribals, encouraging communities to manage their own resources and assisting and negotiating the government in the management decisions. NGOs in the tribal economy have played a facilitator in bridging the credibility gap as well as in providing assistance on the non-technical and social and economic aspects of participatory CPLRs management as distinct from the technical aspects that are more commonly handled by forestry authorities. In the absence of any management institutions, a very few number of tribal households however preferred the individual capacity level of managing CPLRs over community management only due to their poverty and livelihood needs, conflict among the users in the village and depletion of unity, cooperation and coordination among the community members. However, control of large scale forest fire for podu and encroachment and unauthorized felling of trees for sale in the market are beyond the capacity of individual households. For this they strongly recommend the community level management to control large scale podu cultivation, solve water problem and obtain financial assistance.

III. Conclusion and Policy Implications

Poverty of tribals coupled with absence of secured property rights and well-demarcated boundary leads to the tragedy of the CPLRs in the tribal areas and pockets and emergence of various institutions of managing CPLRs. Over time, communal ownership and usufruct rights of the tribals have been restricted and curtailed and even delegitmised in response to penetration and influence of outsiders, market forces, development of transport and communication, establishment of development and mining projects, reservation of forests as protected and reserved category. There is conflict of interest between the tribals and government for the specification of secure and exclusive property rights as the former can not be displaced from their land of habitation and subsistence means of livelihood. With a view to managing CPLR, the government emphasizes nationalization of forests and formulates forest and environmental policies and Acts. In spite of this, government failure persists on account of several reasons of which principal ones are abject poverty of the tribals, non-payment of compensation to the tribal households while introducing afforestation on the patch of land otherwise used by them for shifting cultivation and non-assignment of property rights in appropriate form. Failure of the government to manage CPLRs in the tribal economy has shifted the management regime to either to community-based management or comanagement or both and some times NGO promoted management of CPLRs. Community management fails due to depletion of unity and integrity among the community members, inadequate finance, non-specification of property rights, and water, fencing and guarding problem and ill-defined resource boundary. Therefore, participatory management through people's participation and collective action (JFM, for example) is the most delicate one to deal with the tragedy of the commons. Poverty of the tribals, conferment of property rights and ill-defined resource boundary are one of the constraints to both the management regime. Between the community management and comanagement, most of the tribals favoured the latter due to financial assistance and conferment of property rights. However, both the management regimes require more financial assistance from the government to meet shortage of irrigation, technology and fertilizer used.

On the whole, the present study examines the role and relevance of existing institutional arrangements for efficient management of CPLR in a tribal pocket of Orissa and has arrived at the inference that all such institutions have failed to meet the desired objective in the absence of conferment of property rights on the one hand and well-defined boundaries of operation on the other. Ultimately the study makes some recommendations on the basis of the findings in respect of improving the sustainability of CPLR use in a tribal dominated area. In order to improve management of CPLR in the tribal economy, the indigenous people should be given more freedom to act on their own. The state should provide resources and assistance by formally allowing them a share in the CPLR products. This would enhance the development of tribal areas and pockets and the mutual trust between the user groups so that mutual participation can be sustained by getting closer to the optimal level of social capital. Appropriate compensation measures and coordination between the government and user groups should be required for the effective management of CPLR in the tribal economy. Assignment of secured property rights and breaking of the vicious poverty among the tribals living in the ecologically fragile regions of the sample blocks in Keonjhar district precipitate necessary prerequisite for the efficient management of CPLRs including forests.

REFERENCES

Adhikari, Bhim (2005), 'Poverty, property rights and collective action: understanding the distributive aspects of common property resource management', *Environment and Development Economics*, Vol. 10, pp. 7-31.

Agrawal, A. (2001), 'Common Property Institutions and Sustainable Governance of Resources', *World Development*, Vol. 29, No. 10, pp. 1649-72.

Baland, J-M and J-P Plateau (1996), *Halting Degradation of Natural Resources: Is There a Role for Rural Communities?*, Oxford University Press Inc., New York.

Bromely, D.W. (1991), *Environment and Economy: Property Rights and Public Policy*, Basil Blackwell, Oxford, UK.

Chopra, K., G.K. Kadekodi, and M.N. Murty (1989), 'People's Participation and Common Property Resources', *Economic and Political Weekly*, December 23-30.

Chopra, K., G.K. Kadekodi, and M.N. Murty (1990), *Participatory Development: People and Common Property Resources*, Sage Publications, New Delhi.

Coase, R. (1960), 'The Problem of Social Cost', *Journal of Law and Economics*, Vol. 3, pp. 1-44.

Demsetz, Harold (1967), 'Towards a Theory of Property Rights', *American Economic Review*, Vol. 57, No. 2, pp. 347-59.

Duraippah, A.K. (1998), 'Poverty and Environmental Degradation: A Review and Analysis of the Nexus', *World Development*, Vol. 26, No. 12, pp. 2169-79.

Jodha, N.S. (1986), 'Common Property Resources and Rural Poor in Dry Regions of India', *Economic and Political Weekly*, Vol. 21, No. 27, pp. 1169-81.

Jodha, N.S. (1990), 'Rural Common Property Resources: Contribution and Crisis', *Economic and Political Weekly*, Vol. XXV, No. 26, p. A-65.

Iyengar, S. (1989), 'Common Property Land Resources in Gujarat: Some Findings about their Size, Status and Use', *Economic and Political Weekly*, 24 June, pp. 67-77.

Lise, W. (2000), 'Factors influencing people's participation in forest management in India', *Ecological Economics*, Vol. 34, pp. 379-92.

Marothia, D.K. (ed.) (2002), *Institutionalising Common Pool Resources*, Concept Publishing Company, New Delhi.

Nadkarni, M.V. (2000), 'Poverty, Environment, and Development: A Many-Patterned Nexus', *Economic and Political Weekly*, April.

National Sample Survey (1999), 'Common Property Resources', NSS 54th Round, (Jan.-June 1998), Draft Report No. 452 (54/3.3/31), NSSO, Department of Statistics, Government of India, New Delhi.

Olson, M. (1965), *The Logic of Collective Action: Public Goods and the Theory of Groups*, Harvard University Press, Cambridge.

Ostrom, E. (1990), *Governing the Commons: The Evolution of Institutions for Collective Action*, New York, Cambridge University Press.

Ostrom, Elinor (2000), 'Collective Action and the Evolution of Social Norms', *Journal of Economic Perspectives*, Vol. 14, No. 3, pp. 137-58.

Pasha, A.S. (1992), 'Common Pool Resources and Rural Poor: A Micro Level Analysis', *Economic and Political Weekly*, Vol. XXVII, No. 46, pp. 2499-2503.

Pearce, D. and R.K. Turner (1990), *Economics of Natural Resources and the Environment*, Harvester Wheatsheaf, New York.

Runge, C.F. (1981), 'Common Property Externalities: Isolation, Assurance, and Resource Depletion in a Traditional Grazing Context', *American Journal of Agricultural Economics*, Vol. 63, No. 4.

Sethi, Rajiv and E. Somanathan (1996), 'The Evolution of Social Norms in Common Property Resource Use', *American Economic Review*, Vol. 86, No. 4, pp. 768-88.

Singh, Chatrapati (1994), 'Managing Common Pool Resources: Principles and Case Studies', Oxford University Press, New Delhi.

Wade, R. (1987), 'The Management of Common Property Resources: Collective Action as an Alternative to Privatization or State Regulation', *Cambridge Journal of Economics*, Vol. 11, No. 2, pp. 95-106.

Ecotourism as an Environmentally Sustainable (Development) Process: A Case Study of Sonamukhi Forests (Bankura)

Mala Bhattacharjee and Rajrupa Mitra

Our study makes an attempt to draw a link between environment and sustainable development. Sustainable development has been referred to as "meeting the needs of the present generation without compromising the ability of the future generations to meet their own needs." An obvious implication of this is that—production consumption and demographic patterns ought to be made 'sustainable' in order to ensure healthy and prosperous lives for the future generations. Injudicious or unsustainable developmental activities leave large masses of people poor and vulnerable as also prone to mal-nutrition and health ailments, apart from deteriorating the environmental quality.

The study is based on a survey undertaken in the Sonamukhi forest area of Bankura (West Bengal). The findings reveal many inadequacies in the infrastructure like underdeveloped roads, transport and communication, shortage of power supply, water scarcity and other amenities.

The surrounding forests in the area primarily serve as a source of livelihood for most of the inhabitants. The natural beauty of the region may be fruitfully and effectively utilized for promoting tourism. Consequently, infrastructural facilities will improve and benefit the local people. Moreover, this will automatically generate more employment opportunities and their income-earning prospects. Demand for forest-based products and handicrafts will be boosted up, which in turn, will act as an impetus to the growth of local cottage industries. Such beneficial externalities will lead to multiplier effect of income generation and stimulate the entire economy of the region. Of course, development of tourism shall have to be consistent with the norms of sustainability (referred to as ecotourism) without adversely affecting the environmental quality. Our study highlights the fact that an overall equitable, efficient and environmentally full-proof improvement of the region may be effected through ecotourism which seems to be a viable (and feasible) mode of attaining environmentally sustainable development there.

THE ISSUE OF SUSTAINABILITY

According to the World Commission on Environment and Development (1987), sustainability refers to "meeting the needs of the present generation without compromising the ability of the future generations to meet their own needs". Sustainability, therefore, has inherent in it short-term equity (i.e. meeting present needs) as well as a much deep-rooted concept of inter-generational equity. Such an issue induces the environmental policy-makers to balance the criteria of economic efficiency and equity. Now-a-days, there is an emerging global consensus on the finite and limited carrying/assimilating capacity of the environment. It has also been acknowledged that global consumption, production and demographic patterns ought to be made 'sustainable' in order to ensure healthy and prosperous lives of future generations. Economic development, indeed, demands high priority—however its adverse effects are of much concern. The massive increase in world production and consumption has, no doubt, stimulated economic growth but has also led to rampant, injudicious and unsustainable use of the environmental resources and damaged the environment, apart from wielding adverse effects on the socio-economic lives of the people.

The World Commission on Environment and Development, in its report entitled, "Our Common Future" stated that developmental activities in various countries were having harsh consequences by making large masses of people poor and vulnerable apart from deteriorating the environmental quality. Infact, unsustainable development is reflected through wide scale poverty, mal-nutrition and health ailments of people. Hence the Report stressed that development ought to be such as to sustain human progress not only for the present, but also for the future—and this is precisely where the issue of sustainability assumes importance. In order to rectify the unsustainable modes of development and accomplish sustainability, the production processes must be efficient and waste minimizing, consumption patterns must generate minimum waste, demographic pressures must be reduced and access to health care must be ensured for all. Development must therefore be equitable, efficient, environmentally full-proof as well as consistent with the broad societal goal of sustainability.

The issue of sustainability or sustainable development not only focuses on methods of protecting the environment but also how productive and rational use of natural resources can be made so as to eliminate poverty and improve human welfare and quality of life. Sustainability includes:

- a concern for the future generations (inter-generational equity) and ensure their economic and social viability;
- a concern for human welfare through redressal of poverty;
- involving local communities in planning and developmental projects; and
- wielding a positive impact on environment and its resource use.

The fact remains—in the context of both intra and inter-generational distribution of costs and benefits accruing from developmental activities, the issue of primary concern in the theory of sustainable development is improvement in the quality of life of the people as well as their enhanced access to natural resources and protection of the natural environment.

ENVIRONMENT AND SUSTAINABLE DEVELOPMENT

The resource-base or stock of natural resources of the environment is not inexhaustible and sustainable development can be viewed as economic development if development is undertaken in accordance with the principles of environmental sustainability. Hence, without environmental sustainability it is impossible to achieve sustainable development.

When environmental degradation remains unaccounted for, the measurement of development will be inappropriate. The dual approaches of sustainable development and environment (launched at the United Nations Conference on Environment and Development) stress on the fact that environmental problems ought to be tackled in a manner such that developmental needs of developing countries are not ignored. Economic growth and poverty eradication are the prime concerns of developing countries apart from the pressing environmental priorities. This necessitates adoption of both corrective and preventive measures and integration of issues pertaining to both environment and sustainable development.

Daly (1990) suggested three essential principles for sustainable use of the environment and natural resources:

(i) As far as the physical volume of inputs and outputs are concerned, the overall scale of resource use must be limited and technological progress must be directed towards maximization of efficiency;

(ii) The renewable resources must be exploited on a profit-maximizing, sustained yield basis and thereby prevent their extinction. This implies:
(a) as far as the resources which serve as inputs (e.g. plants and animals) are concerned, the harvesting rate should not exceed the regeneration rate, and (b) as regards, the resources which serve as 'sinks' (e.g. the atmosphere or water bodies), the wastes/effluents generated should not exceed the renewable assimilative capacity of the environment.

(iii) With respect to the exhaustible natural resources, sustainability principle requires maintenance of the total stock of natural capital (by depletion of the non-renewable natural components (e.g. minerals) at a rate which corresponds to the rate of creation of renewable substitutes.

Thus an important aspect of sustainable development has been maintenance, regeneration and restoration of the fragile ecosystem.

Furthermore, it is extremely important that sustainable development must take into account local aspirations because local communities are a part and parcel of their surrounding environment. The necessity of involving local communities into the framework of environmentally sustainable development stems from the fact that rural/local inhabitants possess a much better understanding of their surrounding natural environment, flora and fauna which is passed on from the preceding to the subsequent generation. Hence, it is necessary for the countries to adhere to stringent environmental norms for attainment of sustainable development such that local people are made accountable for ensuring sustainability of the environment—and it is on their efforts that the environmental resources can be mobilized properly and local needs and aspirations can be met with the ecological concern in mind.

Empirical studies have revealed a positive correlation between involvement of local people and attainment of sustainable development. To quote Brown J.W. and Gobaldon A.J. (1993), "like democracy, sustainable development will require widespread civic participation at all levels of decision-making as its cornerstone, and quality participation requires quality education". As a matter of fact, dissemination of information (among local inhabitants in particular) is a pre-requisite for sustainable development—because it has been asserted that it is only through enlightened understanding of the environment by a person that it can be saved. Moreover, knowledge of technologies (which are the quickest routes to sustainability) is also important.

As we have already stated, lopsided and injudicious developmental activities (without any concern for the environment) have accelerated the pace of environmental degradation. Such a form of environmental degradation is manifested through growing scarcity of natural resources and using the natural environment as a 'sink' for wastes and pollutants—the consequence of which is a severe threat to sustainability of the economy. Needless to say, if the sustainability of the environment is endangered, the very existence of the local people will be at stake. This makes it all the more important to generate awareness among the local communities with regard to environment and sustainable development. The Agenda 21 (as finalized by the United Nations Conference on Environment and Development) prescribed "A major priority is to reorient education towards sustainable development"—because it is crucial for achieving such awareness, values, skills and behaviour which are consistent with the idea of sustainable development.

Environmental sustainability requires safe disposal of wastes (from industrial activities), safeguarding the natural ecosystem and curbing rampant and injudicious exploitation of natural resources. Besides, priority is accorded to the poorer/more deprived sections of the community who are compelled to bear the greatest burden of environmental degradation due to modernization, industrialization and other forms of developmental activities.

Environmental sustainability, therefore, incorporates the ideologies of:

(i) bringing about improvement in environmental quality;
(ii) generating public awareness on environmental problems, their solutions and conservation techniques, and
(iii) providing/encouraging people's participation in decision-making such that their ability to evaluate developmental programmes is enhanced.

The very essence of the theory of sustainable development is fulfilment of the basic and minimum needs of the poor—without compromising with the ability of the environment to provide similar benefits to the future generations. In this context, people-environment-economy linkage is stressed upon for protection and conservation of the environment as also for abating various forms of environmental degradation. It is only through a proper and appropriate synchronization of human rights and socio-economic and political aspirations of the people, in conjunction with the norms of environmental protection and conservation, that the principal motive behind environmental sustainability (with simultaneous benefits conferred on the people and environment) can be accomplished.

ECOTOURISM AS A VIABLE AND SUSTAINABLE SOLUTION

The concept of ecotourism is gradually gaining popularity which involves a change in tourist perceptions, increases environmental awareness and involves a desire to explore natural resources and the environment. As per definition and principles of ecotourism provided by The International Ecotourism Society 1990, ecotourism is "Responsible travel to natural areas that conserves the environment and improves the well-being of local people". Therefore, ecotourism has inherent in it the ideal of welfare of the local people residing in that area and this is where the basic principles of ecotourism and sustainable development coincide.

The importance of industrialization in the current era of modernization is unquestionable. Tourism industry has also been growing rapidly in recent years, keeping pace with the development of new industries. India, too, began to accord increasing importance to tourism since the 1990s upon realizing its income-generating potential. In 2002, there was a 14.6% hike in the number of foreign tourists visiting India leading to increased foreign exchange earnings by around 22.4%. Infact, tourism industry is the highest foreign exchange earning industry in India.

One of the fastest flourishing industries of the world is ecotourism industry, contributing about 6% of the GDP all over the world. The year 2002 had been declared as the International year of Ecotourism. An ecotourism industry includes all activities undertaken for maintaining the natural surrounding. It also supports recycling of ecologically safe objects, reuse of water, efficient use of natural resources and also generates employment opportunities for the local people, apart from ensuring that the place (visited by tourists) is not spoiled in any way and ecological or natural equilibrium of the place is not disturbed.

The developing countries are characterized by natural

biodiversity and presence of varied species of wild life and flora and fauna. However, due to rapid population growth, over exploitation of resources, de-forestation and pollution, the developing countries are facing loss of bio-diversity. In this regard, tourism can be applied as an efficient strategy for mitigating the socio-economic problems faced by the developing nations (both at regional and national levels) and enable them to conserve their precious biodiversity and natural resources.

The essential principles embedded in ecotourism are:

- to minimize impact on the environment
- to build environmental and cultural awareness
- to provide positive/beneficial experiences for both visitors and hosts
- to provide financial benefits and thereby empower the local people.

Some suggestions or inputs for community-based ecotourism (i.e. which involves and benefits the local communities) are:

- Provision of guide services to the visitors
- Setting up sale outlets for local products and souvenirs.
- Promotion of recreational facilities like boating, fish angling, folk dance, picnic spots, nature trails, etc.

As far as the effects of ecotourism are concerned, there are 2 conflicting ideas referred to as the optimistic view and the pessimistic view. As per the optimistic view, tourists are looked upon as an economic force promoting conservation of natural resource. The revenues earned from tourist visits (e.g. entrance fees, domestic air fares, accommodation and food, hiring charges of guides, etc.) get distributed among the people of the home country. Besides, more complex economic linkages also exist which cause transfer of resources from people selling goods and services to the tourists to other members of the local economy who sell goods and services to the agents. For example—hotels hire local workers, pay rents to locals and also purchase local "intermediate inputs" like fruits and vegetables, fish, meat, etc. Payments made for such goods and services also influence the incomes of the local people. These people again stimulate new rounds of expenditures which again influence incomes of still more people thereby leading to a multiplier effect in income generation.

The pessimistic view about ecotourism is that—sustainable development can not be achieved by ecotourism. Tourism has been placed as the second major threat to protected areas by the IUCN (1992). Distribution of benefits accruing from tourism is another problem. Conservation of ecologically fragile and protected areas entails opportunity costs and hence if the beneficial impacts are not fairly distributed, it will cause severe loss of welfare. Impact of tourism on the environment, society and economy are multidimensional and quite complex. Since ecotourism implies greater demand for the pristine un-degraded natural areas, higher pressure is exerted on the fragile ecosystem. Critics opine that green-washing practices (which are often carried out in the name of ecotourism) are instances of harmful/undesirable effects of ecotourism. Such practices are referred to as commercialization of tourism schemes. An example of green washing may be construction of a hotel in a natural, scenic landscape which considerably affects the natural ecosystem. Other problems associated with ecotourism are—it diverts resources from projects which might contribute to more sustainable and realistic solutions of pressing social and environmental problems; it causes conflicts over land, resources and tourism profits and sometimes fails to confer benefit to the local communities, etc.

Ecotourism also leads to consumption of virgin/unspoiled natural territories, deforestation, disruptions of ecological life system, violation of the ideals of conservation and generates various forms of pollution. In many cases, profits accruing from this industry do not reach the local people—but are reaped by investors investing in this industry. Local people might not tolerate visits by tourists beyond a particular limit. As a solution to this problem, the carrying capacity of a particular region/tourist spot must be ascertained and tourism beyond this limit should be strictly prohibited.

Thus, expansion of tourism leads to pressures on the local population. There is also large scale migration to areas of tourist interest in search of jobs leading to increase in population in those areas. Tourism also imposes high costs on the environment and degrades the environment considerably. Tragedy of the commons poses an example economic un-sustainability from environmental protection. Although the community, as a whole, has incentives to protect the environment, companies undertaking ecotourism exploit the ecotourism site unsustainably. Lack of environmental sustainability raises need for small scale, slow-growth and locally-based ecotourism. Since the local people have a vested interest in the well-being of their community, they are more accountable for environmental protection. If the local communities make increased contributions towards locally managed ecotourism, viable economic opportunities may be created and environmental issues associated with poverty and unemployment may be properly tackled.

Even if ecotourism might not always be ecologically or socially beneficial, its importance as a strategy for conservation and development can not be denied. Inspite of doubts expressed by critics it can indeed be asserted that benefits of ecotourism are quite large. Some such benefits are:

- By involving local communities, ecotourism leads to economic development of the local area.
- It strengthens the interface between the ecotourists and local people.

The negative side of ecotourism is when it is viewed as a trade, local people may not be empowered and are often exploited. Hence, for ecotourism to be useful and fruitful to local communities, it must be re-oriented and made sustainable. Only then can ecotourism be looked upon as a viable and sustainable process of development.

Ecotourism may be developed through natural resource management. Natural resources are fast depleting with increases human encroachment and habitats—and ecotourism programmes should be modified with a view to conserve natural resources. Nevertheless, it ought to be emphasized that the type and scale of development of ecotourism should be compatible with the environmental and socio-cultural characteristics of the local community. Moreover, ecotourism should be planned and undertaken as an inherent part of the overall strategy for development of the area. It has, therefore, been asserted that stringent laws and regulations should be implemented to check the spread of unsustainable ecotourism. For sustainable growth of ecotourism effective planning and management are of vital importance which will ensure that development of tourism is undertaken, keeping in mind the socio-economic priorities of the local people and overall environmental and ecological well-being.

Forests as a Means of Environmentally Sustainable Development

Forests constitute a very precious, renewable natural resource and an important component of the ecosystem. Unsustainable developmental processes have been reflected through deforestation, reduction in forest cover and ultimate desertification in many areas. Multi-pronged pressures have been exerted on the forest ecosystem through growing population, cattle grazing, fuel and fodder collection, industry and forest fires. These amply demonstrate the impact of unsustainable development on forests. As regards India, forest cover is estimated to be only 11% whereas the desirable forest area should be 33% of the total land area (as specified by the National Forest Policy). It may be stated that till the late 1970s, forests were targeted for re-settlement, agriculture and industrialization. Such a trend was considerably checked by the Forest Conservation Act of 1980. Of late, however, there has been the emergence of wide scale consensus on the ill-effects of deforestation and hence policies are being adopted at all levels for restoration of this precious resource: The importance of forests in sustaining various species of life (both plants and animals), and thereby maintaining biodiversity is unquestionable. Steps are also being taken in the direction of promoting afforestation and increasing the area under forests.

Forest cover may be increased by:

- Improving canopy cover in the forest land
- Undertaking large scale afforestation, particularly in the non-forest and degraded lands.

Although steps taken in this direction have yielded positive results, we are yet to achieve the desired extent of forest cover/ afforestation.

In the state of West Bengal the tree cover is 30.74% of the geographical area as per the satellite imagery survey carried out in 2006. Under the National Forest Policy (NFP) 1988, forest conservation was stressed. During 2006-07, in West Bengal, a total plantation of 18571.3 hectares has been raised through afforestation programmes. Joint Forestry Management has been successfully implemented in the state. Till 31st March 2007, 4049 Forest Protection Committees (FPCs) and 102 Eco-Development Committees (EDCs) have been created. These FPCs and EDCs are protecting a total forest area of 631611 hectares.

We present, in the table below, the forest cover in the state of West Bengal in 2003.

TABLE 1

Forests in West Bengal (Sq Km): 2003

Geographical Area	*Actual Forest Cover*	*Dense Forest*	*Open Forest*	*Mangrove*	*Scrub*	*Non-forest*
88752	12343	6045	6298	2120	75	76334

Source: State of Forest Report, 2003, as quoted in Statistical Abstract, p. 84.

West Bengal is quite rich in a variety of forest products. We present below the output of forest produce in West Bengal during 2000-01 and 2001-02.

TABLE 2

Forest Produces in West Bengal

Year	*Timber (cubic metres)*	*Fuelwood (cubic metres)*	*Bamboo (nos.)*	*Other NTFP (tonne)*	*Sal seeds (tonne)*	*Tendu/kendu/ Bidi Leaves (tonne)*
2000-01	49047	189735	89 NT	2291	1848	371
2001-02	22853	270729	119 NT	7399	—	—

Source: Forestry Statistics India, 2003, as quoted in Statistical Abstract, pp. 90-93.

Production of timber and firewood from the forests of West Bengal is quite substantial. In the table below we present data on the out turn of timber and fire wood from forests in West Bengal.

TABLE 3

Timber and Firewood Production (in cubic metres) from Forests of West Bengal

Year	*Timber** *(in cubic metres)*	*Fire wood*** *(in cubic meters)*
2002-03	102357	218469
2003-04	130551	306729
2004-05	113871	366583
2005-06	85993	324092
2006-07	114589	387094

*Timber includes pulpwood, matchwood, other timber and poles.

** Firewood includes quantity of firewood required for production of charcoals, other firewood and pulpwood.

Source: Directorate of Forests, Government of West Bengal—as quoted in *Economic Review*, Statistical Appendix, 2007-08, p. 97.

The formation of the West Bengal Forest Development Corporation Limited in 1974 marked a change in the situation of forestry in West Bengal. Its main objective was to promote and develop forestry and wood industry. In the year 2005-06, the West Bengal state government fund was Rs. 552.75 lakhs as shares. Other sources contributed Rs. 70 lakhs as shares and Rs. 500 lakhs as loan. The net profit of the West Bengal Forest Development Corporation Limited was Rs. 1214.57 lakhs at the end of the year 2005-06. This obviously was a positive step towards the promotion of forestry in West Bengal.

We show the revenue and expenditure accrued to the Government of West Bengal from and on forests in the year 2001-02.

TABLE 4

Revenue and Expenditure from/on forests in West Bengal

(*Rs. Thousands*)

Revenue	*Expenditure*	*Net Revenue/Deficit*
263952	1462073	–1198121

Source: Forestry Statistics India, 2003, as quoted in Statistical Abstract, p. 94.

Moreover in the Table 5 we show only the revenue receipts accruing to the West Bengal Government from forests.

TABLE 5

Revenue Receipts from Forests (West Bengal)

(*in Rs. Lakhs*)

Year	*Forest Directorate*	*Forest Development Corporation*	*Total*
2002-03	5622.34	5067.10	10689.44
2003-04	4463.10	5406.18	9869.28
2004-05	3986.79	6534.14	10520.93
2005-06	3745.28	7648.53	11393.81
2006-07	4044.88	8866.75	12911.63

Source: Directorate of Forests, Government of West Bengal and West Bengal Forest Development Corporation Limited—as quoted in *Economic Review*, Statistical Appendix, 2007-08, p. 98.

The above tables reveal that revenue earned by the government of West Bengal from forests is quite substantial.

In the following table, we present data on wastelands in the districts of West Bengal.

TABLE 6

Wastelands of West Bengal

No. of districts covered	*Total geographical area of districts covered (sq.km)*	*Total wasteland area in districts covered (sq.km)*	*Percentage of total geographical area*
18	88752.00	5718.48	6.44

Source: Forestry Statistics in India, 2003, as quoted in *Compendium of Environment Statistics*, p. 52. Since the percentage area of wasteland is substantial they can be put to productive use.

A SOCIO-ECONOMIC PROFILE OF THE CASE STUDY AREA

Since our case study was conducted in the district of Bankura we shall now try to highlight on the district level statistics of Bankura, before coming to the specific location of our case study viz, Sonamukhi block of Bankura. The geographical area covered by Bankura district is 6881 sq. km. with a total population of 31,92,695—the density of population being 464 per sq. km.

TABLE 7

Percentage of Land Utilization—Bankura and West Bengal

Place	*Reporting Area*	*Area under Forest*	*Area not available for cultivation*	*Other uncultivated land excluding current fallows*	*Current fallows*	*Net area sown*	*% of cultivable area* to total area*	*% of net area sown to cultivable area*	*Cultivable area* per agricultural worker(a)*	*Net area sown per agricultural worker(a)*
Bankura	688.00	21.68	21.54	0.81	5.88	50.09	55.97	89.50	0.56	0.50
West Bengal	8684.11	13.52	20.20	1.37	3.93	60.98	64.91	93.95	0.52	0.49

* = Current fallows + net area sown.

(a) = Agricultural workers (Agricultural cultivators + Agricultural labourers) main and marginal as per census 2001.

Source: *Economic Review*, 2006-07, Statistical appendix, pp. 212-13.

Primarily the economy of Bankura is agrarian in character, agriculture accounting for almost 70% of the income of the district.

We compute the percentage figures for the different types of land utilization in Bankura and West Bengal in 2006-07. (Table 7)

Although Bankura has reasonable forest cover there are fallow lands lying idle. If these lands are put to cultivation or used in any other form of income generating activity economic position of the area can be improved.

Bankura is characterized by tropical, dry deciduous forests. Amount of dense forests have been considerably reduced in Bankura and many shrubs have become extinct. Even then, it has a variety of trees, shrubs and creepers. The deep forest areas in Ranibandh, Khatra, Sonamukhi, etc. are rich in natural, rare and herbal plants. The trees in the forest are used as raw materials in paper and tobacco industries, for making furniture, as fuel, and put to medicinal uses by the local people. Flora species of the forest include the most characteristic varieties like Sal, Bahera, Piasal, Kend, Palash, Mango, Jam, Haritaki, Sisu (Indian rosewood), Neem (Margora tree), Siris (Parrot tree), Amlaki (Embeli), Bel (Wood-apple), etc. Apart from these, there are other creepers and/or shrubs like Basak, Shatmuli, Bera, Kalmegh, Anantmul, Bantulsi, Thankuni, Kantiikari, Kulekhara, Somraj and Hastikarna. In Bankura a significant percentage of forests are protected forests, i.e. where all activities are permitted unless prohibited. We show the area under forests (by legal status) in both Bankura and West Bengal.

TABLE 8

Area under Forests by Legal Status in Bankura and West Bengal, 2006-07

Place	*Reserved Forests (sq. km.)*	*Protected forests (sq. km.)*	*Unclassed forests (sq. km.)*	*Total forest area (sq. km.)*	*% of total forest area to geographical area*
Bankura	80	T311	91	T482	21.53
West Bengal	7054	3772	1053	11879	13.38

Source: Economic Review, 2007-08, Statistical Appendix, pp. 212-13.

The percentage of total forest area to geographical area is higher in Bankura than in the state of West Bengal. Such forests apart from providing valuable forest-produces also earn revenues and serve as attractive tourist spots.

We now concentrate on our case study area viz. Sonamukhi. Sonamukhi and Sonamukhi Municipality (M) are within Bishnupur Sub-Division of Bankura district. We take a look at the area, population and density of population in these areas, which we present in the table below.

TABLE 9

Area and Population of Sonamukhi Forest Area (Bankura)

Sub-Division/ Block	*Area (sq.km.)*	*Population*	*Population density (per sq.km)*	*% of population to district population*
Bishnupur S.D.	1870.05	996348	533	31.20
Sonamukhi	378.85	142328	376	4.46
Sonamukhi (M)	11.65	27354	2348	0.86

Source: Census 1991, 2001.

The literacy figures (as per 2001 census) for Sonamukhi are presented in the Table 10.

TABLE 10

Literacy Rates—Sonamukhi

Place	*Male*	*Female*	*Total*
Sonamukhi	70.7	45.9	58.7
Sonamukhi (M)	86.5	70.7	78.8

Note: Literacy excludes children in the age group 0-6 years.
Source: Census 2001 and District Statistical Handbook, 2005.

Apart from literacy an important indicator of the well-being of an area is the availability of health care and medical facilities, which we depict in the Table 11.

TABLE 11

Health Care and Medical Facilities in Sonamukhi

Place	*Hospitals*	*Health Centres*	*Clinic*	*Dispen- saries*	*Total beds*	*Doctors*	*Family welfare centres*
Sonamukhi	—	4	23	1	22	04	23
Sonamukhi (M)	I	1	3	4	30	07	—

Source: District Statistical Handbook, 2005.

We now present different types of data on social and economic aspects of our case study area viz. Sonamukhi. This will enable us to have a vivid overall picture of the block.

TABLE 12

Sonamukhi: Socio-Economic Profile

12A: Common Facilities

No. of mauzas having drinking water facility	*No. of fair price shops*	*Gram panchayat office with phone*	*No. of fertilizer depots*	*No. of seed stores*
160	53	10	57	8

12B: Commercial and Gramin Banks

No. of commercial bank	*No. of gramin bank*	*Population served per bank office ('000)*	*No. of cooperative societies*	*No. of members*	*Working capital*
2	3	28	50	12271	74446

12C: Roads in Sonamukhi (km)

P.W.D.		*Zilla Parishad*		*Gram panchayat samity*	
Surfaced	*Unsurfaced*	*Surfaced*	*Unsurfaced*	*Surfaced*	*Unsurfaced*
44.00	10.00	23.40	28.00	—	440.00

12D: Transport Facilities

No. of ferri service	*No. of originating/ terminating bus route*	*Distance of nearest railway station from block head quarters (km)*
4	10	25

Source: *District Statistical Handbook* (Bankura), 2005.

Furthermore, agriculture being an important income earning occupation, it is necessary to look into data pertaining to agriculture. The table below shows the number of farmers in Sonamukhi. (Tables 13 and 14)

TABLE 13

Number of Farmers

Bargadars	*Patta holders*	*Small farmers*	*Marginal farmers*	*Agricultural farmers*
10173	9130	4519	11119	30374

Source: District Statistical Handbook (Bankura), 2005.

In view of the importance of agriculture in the economy of the region, irrigation facilities are necessary. The table below shows the various modes of irrigation applied in the region viz. canal, tank, lift irrigation etc. (Table 15)

Apart from cultivation, fisheries are also undertaken in the region. In the Table 16 we present data on fisheries of Sonamukhi. (Table 16)

We also depict in the Table 17 the number of livestocks in Sonamukhi.

The statistics on Sonamukhi depicts that for improvement in the standard of living of the local people infrastructural facilities must improve. Irrigation facilities are quite limited thereby hindering cultivation. Consequently crop yield is not much. Water scarcity, apart from affecting agriculture, also affects pisciculture. Hence, many people have adopted livestock breeding as their occupation. The scope of adopting other occupations is also limited. Therefore, for enhancing income earning capabilities of the local people the only route is employment generation. Ecotourism is a viable process in this regard which can upgrade the conditions of the people through generation of employment and inclusive growth strategy. As the region is naturally beautiful, ecotourism can be easily promoted through provision of the necessary infrastructure.

TABLE 14

Crop Yield

Aus			*Aman*			*Boro*			*Jute*			*Barley*			*Whea*		
area	*prodn.*	*yield*	*area*	*prodn.*	*yie ld*	*area*	*prodn.*	*yield*	*area*	*prodn.*	*yield*	*area*	*prodn.*	*yield*	*area*	*prodn*	*yield*
26.0	80.3	3084	128.5	405.2	31 53	56.1	129.8	2315	0.2	2.8	17.8	–	–	–	4.3	10.1	2365

Area is given in '00 hectares, prodn. in '00 metric tons and yield in kg/hectare.
Source: *District Statistical Handbook* (Bankura), 2005.

TABLE 15

Irrigation

Canal	*Tank*		*R.L.I.**		*D.T.W.***		*S.T.W.****		*O.D.W.*		*Others*		*Total*	
Area	*No.*	*Area*	*No.*	*Area*	*No.*	*Area*	*No.*	*Area*	*No.*	*Area*	*No.*	*Area*	*No.*	*Area*
8600	20	180	10	179	65	440	3800	4700	13	15	15	252	3923	14366

Area is given in hectares. * River lift Irrigation. ** Deep Tube Well. *** Shallow Tube Well.
Source: *District Statistical Handbook* (Bankura), 2005.

TABLE 16

Fisheries in Sonamukhi

No. of government schemes operated	*Expenditure ('000 Rs.)*	*Assistance to needy fishermen*	*Net area available for pisciculture (hectares)*	*Net area under effective pisciculture (hectares)*	*No. of persons engaged*	*Approx. Annual production*
3	20	–	1309	1170	6390	16970

Source: District Statistical Handbook (Bankura), 2005.

TABLE 17

Live Stocks

Cattle	*Buffalo*	*Goat*	*Poultry birds*	*Sheep*
80618	8813	50465	641556	355

Source: District Statistical Handbook (Bankura) 2005.

EMPIRICAL FINDINGS FROM THE SURVEY

We now present the findings of our survey conducted in the two villages of the Sonamukhi forest area viz. Kalaberia and Bankalyanpur. In Kalaberia we surveyed almost 260 people (of which 60% were males and 40% were females) and in Bankalyanpur we surveyed 10 tribal families.

Kalaberia

The village is located very close to the forest. Although there is not much interaction between the forest department and the villagers, the villagers are definitely benefited due to conservation of the forest. The government has provided saplings to the villagers for encouraging afforestation. The villagers have received Rs. 3187 (by selling trees which are cut-off). This amount is 25% of the entire revenue of the government after selling wood.

Forest acts as a source of livelihood for most of the villagers. The main activity of the villagers is to collect sal leaves and fuel wood from the forest. The sal leaves are collected by women, in fact, completely and solely performed by them. Selling of Kendu leaves (for bidi-making) is another economic activity. Besides, selling of Mahua fruit, collected from the forest, is another source of earning. However, due to lack of planning, gums from sal tree can not be extracted and used as a means of livelihood or earning. The rates of various forest products prevailing in the markets of nearby areas are quite low e.g. seeds of Mahua flower is Rs. 10 per kg, cost of fuelwood is Rs. 50 per 10 bundles, 70 to 80 sal leaves cost Rs. 6, cost of 1 bundle kendu leaves is Rs. 20 (in some areas this rate is between Rs. 10-12 per kilo), cost of mushroom is Rs. 50 per kg, a bundle of wood costs Rs. 20. This obviously implies that when these products are sold by the villagers for sustenance, their monetary benefit is quite limited.

Agriculture is undertaken by the villagers. But its scope is limited due to many constraining factors. The major problem is lack of adequate water. There are no reservoirs to store water for cultivation—nor are there any means for rainwater harvesting. Moreover, the farmers lack credit facilities, seeds, fertilizers, etc. It is evident from interrogating the villagers that if irrigation water is made available or, in other words, if adequate water availability can be ensured many of them are willing to adopt cultivation as their occupation. The Panchayat does not provide any monetary help to the villagers for undertaking cultivation. The villagers are willing to cultivate (kankur) plant to earn money but have not received any help from the government in this regard. Lack of water is a hindrance to fisheries and aquaculture also. Livestock breeding is resorted to because cultivation suffers from lack of irrigation facilities.

Women are aware of the benefits of uniting themselves into Self Help Groups (SHGs). Most of them are engaged in SHGs. They actively participate in micro-financing activities. They are interested in participating in various self-employed or group employed occupations or cottage industries, provided finance and training are made available to them. Women are aware of family planning. They are interested in educating their children and want them to be properly employed. Medical assistants from the health centres come to the village to give vaccination to the children. Anganwadi education is imparted to the children and meals are provided to them under this scheme.

There is a primary school at Jamdoba where the children get admitted. The SHGs help in preparing mid-day meals for the school children. In Jamdoba School there are classes from class I to IV, about 50-59 students and two teachers. A positive aspect that has been noted is drop out rate is negligible among the children. The school is more or less well-equipped. There is a library in the school, apart from drawing boards and display boards. Health check up of children is undertaken regularly at intervals of every 3 months. However, there are no high schools in the region. Children have to go far away from their homes for higher education. This, infact, acts as a serious deterrent to the girls pursuing higher education. Thus they are compelled to leave their studies—and their parents are left with no other option but to get them married off.

As regards the basic amenities, the villagers complained about the lack of safe drinking water, lack of medical help, lack of high schools, power-cut, etc. The nearest hospital is almost 4 km far. Consequently, the villagers are unable to get immediate medical help when needed. Although electricity is available in most of the households, there are severe power-cuts in the region. School is also not close by and this hinders many children from pursuing studies. Tubewells are the major source of drinking water. There are open wells as well. Consequently, water borne diseases are quite common in the area. There is no industry in the village and the requisite infrastructure for development of industries is unavailable. Furthermore, the region being water scarce and devoid of connecting roads and other infrastructure, employment opportunities for local youths are not much. Hence there remains the necessity of opening up new areas of employment for them to make them economically and financially independent.

Bankalyanpur

Bankalyanpur is a village surrounded by forests. Though the distance between Kalaberia and Bankalyanpur is only 2 km, yet the people of Bankalyanpur (which is a tribal viz. santhal inhabited area) have less access to various services required for their smooth and healthy living. The main activity of the people of Bankalyanpur is to collect and sell sal leaves after weaving them into plates and bowls. The activities of the villagers are confined within the villages where they plough agricultural fields belonging to others (mostly santhals plough the fields of Telis residing close by). They are also engaged in construction of roads and canals whose payments are very little. The other activities of the villagers include bidi-making from kendu leaves and making wine from mahua flower.

Although, women are not educated they have constituted SHGs. They give Rs. 20 per month as their contribution to the SHGs. They have accumulated Rs. 7000 in a bank at Paachal. Their intention is to use the money for productive purposes. They are also interested in taking trainings for undertaking various self-employed occupations. This is very important and necessary for making them financially independent as well as for their source of sustenance. This will also aid in the revival of small scale or cottage industries. What is necessary is the evolution of entrepreneurial class (either private or government) who would train the women to cultivate the artistry and creativity embedded in them.

Bankalyanpur does not have any health centre. Health Centre is at Paachal which is quite far-off. The villagers have to take hired cars form Icharia or Kalaberia to go to hospitals which obviously becomes very expensive for them. However, the children are given vaccinations. Polio is given to children less than 5 years old. There is a primary school in the area but no high school. The villagers are not much educated. Education level is till class VIII to maximum class IX standard.

Bankalyanpur also suffers from several infrastructural deficiencies. There is no electricity. The village is water scarce—water is not available for irrigation. Consequently agriculture suffers. There is no hand pump. Rainwater is the only source of water for cultivation. There are no reservoirs for storing rainwater. There is no sanitary system in the village. Open wells are the only source of drinking water, hence, water-borne diseases are very common among the villagers. The panchayat has not been able to develop sewage and sanitary system in the village. The unhygienic way of living is polluting the entire environment. Communication is very poor. There are no telephone booths. Villagers do not have mobile phones. Connectivity of the village via road networks is poor. Roads are under developed. Rationing system is substandard. The villagers have to suffer a lot to collect kerosene oil which is very scarce. There being no television or radio, the village remains quite secluded. The Santhals generally do not go out of the village in search of jobs. The problem remains—there are virtually no job opportunities for them in village. A major problem of the santhals is the infiltration of the elephants from the forest into the villages destroying the houses, crops and even lives of the villagers.

Comparison Between the Two Villages (Kalaberia and Bankalyanpur)

In the village of Kalaberia living conditions are better than Bankalyanpur because there is electricity, good telecommunication system and health centre (although it is quite far). Most women belong to SHGs and are quite active. In both the villages the common problem is underdeveloped roads and transport systems, lack of irrigation system and acute scarcity of water. As a consequence cultivation is severely hampered. Drinking water is also scarce, wells being the only source serving such purposes. Water borne diseases crop up every rainy season. There are no facilities for rain water harvesting. Although the panchayat has intentions to construct canals to facilitate irrigation, this has not materialized yet.

The government has constituted forest protection committee to protect the forests. However, efforts to redress the problems of the villagers have been quite meagre. The women SHGs want to encourage and promote education among women and generate employment opportunities for them via various training facilities.

The habit of banking among the villagers has been noted in Kalaberia though it is limited to around 10 households only. Lack of access to education (both formal and informal), health-care facilities and employment opportunities are the constraining factors in both villages. Rigidities in the form of caste barriers are also present. An example may be—in Kalaberia the women belonging to gėneral castes do not go to the forests to collect sal leaves because it is considered to be beyond their customs, only the women belonging to scheduled castes perform this activity. The Forest Management Committee is entrusted with the task of keeping vigilance over the forest and protection/preservation of forests and forest resources. However, the forest department does not in any manner promote the employment opportunity or ways of earning of the villagers (many of whom depend solely on the forest products for their living). At the time of cutting old trees (katni), the government pays 25% of the revenue to the villagers for welfare and improvement of the village. At least one member of each family in each village is a member of the forest protection committee.

Relevance of Ecotourism in the Case Study Area

From the findings of our case study we can infer that tourism, to be specific ecotourism, can be developed in the Sonamukhi forest area of Bankura. We feel that if ecotourism (in accordance with the principles of sustainable development) can be promoted in the region—as nature has abundantly bestowed pristine beauty of forests in the region—improvement in socio-economic conditions of the villagers can be brought about. Ecotourism will generate a series of beneficial, positive externalities because prior to its commencement infrastructural facilities shall have to be developed. Connectivity of the region via construction of roads and means of transport, improved communication system, adequate water availability, and uninterrupted power supply are all essential for developing ecotourism—and which in turn well serve to boost up the local economy and well-being of the people.

Obviously all these have to be accomplished with the minimum harm imposed on the environment. Thus improvement in economic/financial status of the people can be achieved without adversely affecting the environment in any way. Ecotourism is basically an area of convergence of ecological well-being and sustainable development.

Development of tourism will generate employment opportunities for the local people in tourist lodges, (as tour guides, etc.) and also positively boost up incomes of those engaged in agricultural production, pisciculture, livestock breeding, etc.—thereby increasing their incomes considerably. It will also render support to local handicraft production and cottage industries by enabling the local people to market their products to the tourists. Proper channelization of the marketable handicrafts need proper avenues—ecotourism can be one way to develop these avenues. Furthermore, promoting 'spots' of tourist attraction will generate revenues, thereby contributing to the wealth and prosperity of the villages. Sustainable development necessitates gradual alleviation and ultimate eradication of poverty. However, as has been noted from our findings the scenario in the two villages is far from being sustainable primarily due to acute poverty and absence of the basic amenities necessary for decent living standards. Thus, it may be asserted that development of ecotourism will generate a (multiplier) effect on the livelihoods of the local people and improve their incomes and standard of living. Ecotourism can therefore be viewed as a remedy to relieve the people of their 'unsustainable' modes of living and ensure all-round socio-economic development of the region, keeping conformity with the norms of sustainable development.

CONCLUSION

Tourism has its testimonies in the earliest records of history and has recorded phenomenal growth during the last century. From 1980 to 2000, global tourism receipts increased at an annual rate of nearly 8%, much faster than the rate of world economic growth of around 3%.

According to Aristotle, work is something which one does for something else (livelihood). Play was mere reaction for work. He deemed leisure as higher and nobler. The concept of leisure has a great influence in determining the mood of mass tourism. To understand the rationale for third world tourism, it is important to distinguish between developed and under-developed countries.

In the developed countries industrialization has been accompanied by absorption of the surplus rural population into the expanding urban industrial centres. This also helped structural reorganization of the rural production sector in the wake of industrialization. In the developing countries the urban industrial sector has not developed sufficiently even to fully absorb the growing urban labour force. A familiar feature of labour surplus third world economy is that in the process of technological progress a significant part of the labour force fails to get involved in gainful employment in the organized sector.

From the economist's point of view, the pertinent question is not whether tourism is socially useful, but whether the resources used for developing tourism have a more useful way of utilization with a comparative advantage coming in the benefit side of the balance. Tourism not only increases the possibility of absorbing more readily available skilled or semi-skilled working force, it also makes better and economic use of certain other resources with low opportunity cost. As tourism develops hotels, motels and resting places, local folks occupy the lowest rank of employment as waiters, room maid, gardener, kitchen helper, porters and so on.

In West Bengal, less industrial development has led to more unemployment or so to say improper utilization of human resources. Recession in the economy has also led to staggering unemployment problem. The survey carried out has been able to highlight certain avenues which need special care. The mass unemployed, yet educated young population of the two villages can be a major source of human resource for the state. The forests and agricultural fields can foster better lives of the villagers if the government shoulders the responsibility of undertaking ecotourism in such a place. The demand management policies (aimed at increasing agricultural production and employment) can go hand in hand with the welfare and upgradation of the local people if the process of ecotourism is properly implemented. Ecotourism will enhance incomes of the people who have no income earning capability/possibility other than agriculture or forestry. The small scale cottage industries need a boost and their marketing can find new ways if tourists visit the places.

Ecotourism will solve the problems of infrastructure like roadways, electricity, water crisis, etc. Moreover, the alternative employment and income-earning opportunities generated by multiplier effect of tourism can be utilized for community welfare. The extent to which the revenue from tourism will be utilized depends on the planners and policy-makers. But so far as the local people are concerned, self-help groups are quite active. So, if the local people are given the responsibility of aiding in the process of development of ecotourism in their locality, apart from overall welfare of the region their worth as human resources can be realized.

Finally, it may be stated that the socio-economic profile of the Sonamukhi forest range is just one example of the entire forest area of West Bengal. If properly analyzed and surveyed, many villages adjacent to forests shall be found to be ideal for ecotourism to develop. It is therefore necessary on the part of the government to promote ecotourism as an alternative source of earning for the people residing in the forest zones of West Bengal.

REFERENCES

Central Statistical Organization (2006), *Compendium of Environment Statistics: India*, Ministry of Statistics and Programme Implementation, Government of India.

World Bank (2008), *World Development Indicators*.

Central Statistical Organization (2007), *Statistical Abstract India: 2005-06*, Ministry of Statistics and Programme Implementation, Government of India.

Bureau of Applied Economics and Statistics (2007), *Economic Review, 2006-07*, Development and Planning Department, Government of West Bengal (with Statistical Appendix).

Bureau of Applied Economics and Statistics (2005), *District Statistical Handbook, Bankura: 2005*, Government of West Bengal.

G.S. Haripriya: *A Note on Eco-tourism* (from website).

www.Bankura.org.

Mohua Guha and Aparajita Chattopadhyay, *Environmental Education: A Pathway for Sustainable Development* (from website).

Jayant Malhoutra (1999), *Environment and Sustainable Development*, 2nd Committee (Economic & Financial) of the United Nations (from website).

Patricia Kameri-Mbote, Philippe Cullet (1996), *Environmental Justice and Sustainable Development: Integrating Local Communities in Environmental*

Management, Working paper of International Environmental Law Research Centre (from website).

Vennila Govindaswamy, Importance of Environmental Education for Sustainable Development (from website).

Bureau of Applied Economics and Statistics (2008), *Economic Review, 2007-08 and Statistical Appendix,* Development and Planning Department, Government of West Bengal.

Website of West Bengal Tourism Department, www.westbengaltourism.gov.in.

Website of West Bengal Forest Development Corporation, www.wbfdc.com.

www.westbengalforest.gov.in.

T. Panda, S. Mishra and B.B. Parida (eds.) (2004), *Tourism Management: The Socio-Economic and Ecological Perspective,* Hyderabad: University Press (India) Pvt. Ltd.

Kunal Chattopadhyay (2008), *Understaning Tourism Economics,* Kanishka Publishers, New Delhi.

Ecological Approach to Remove Chronic Poverty and Sustainable Development for India

Somnath Hazra and Raj Kumar Sen

Under the phase of economic reforms in the low income countries as well as throughout the world, the number of people living in chronic poverty has increased during the last five years. In this period the number of chronically poor people lies on a range between 320-443 million. The characteristics of chronic poor people not only depends on low income but also on deficiencies like hunger, malnutrition, illiteracy, lack of safe drinking water and primary health services, social discrimination, lack of physical security and political exploitation. Here poverty, environment and development have a complex relationship in the developing countries. Since 1991, growth rate is increasing from 5.4% to 8% and Below Poverty Line (BPL) population has reached a level of 26.7% in India.

Based on the earlier draft of the paper accepted for presentation in the International Conference *"Eradicating Chronic Poverty in India: Policy Issues and Challenges"* (1-3 October, 2008) held at Jawaharlal Nehru University, New Delhi.

In the last century the global population increased from 1.6 to 6.1 billions. The massive population growth leads to slow down of per capita income growth due to resource constraint, and degrading environment due to urbanization, industrialization and expansion of agricultural land. After the reforms the poverty reduction policies are paying attention on two basic goals: (1) achieving macroeconomic stability to strengthen economic growth, and (2) providing the basic factors (health, education, water and sanitation) for human development.

In this paper we have analyzed the nature of chronic poverty and we are trying to find out causes of failure to eradicate it from India with emphasis on the faulty official policy and the massive corruption in funding for poverty alleviation. Even with some heterogeneity of chronically poor people, we have identified some main issues that strengthen chronic poverty, like insecurity, political voicelessness, physical remoteness, social inequity, exploitation, etc.

The poorer people are suffering from the ill-effects of the skew consumption pattern where 30 percent of the world's rich population consumes 70 percent of the world's resources which leads to less accessibility of resources for the poor people who are basically dependent on natural resources for their survival. A few case studies involving ecological poverty and their alleviation are given as observed in Ralegan (Maharashtra), and Arabari Range (West Bengal) etc.

Poverty alleviation and environment protection are mutually reinforcing and support economic growth. Environmental risk factors are major source of health problems in developing countries. Hence environmental degradation produced unsustainable development and alleviation of poverty is the determining factor of sustainable development. Poverty alleviation is no longer possible with continued dependence on the trickle down effect of economic growth. A focus on ecological poverty may show the right direction.

Key Words: Chronic Poverty, Ecological Poverty, Sustainable Development. JEL **Classification:** P36, P46, Q01.

I. Introduction

While the frontiers of poverty analysis have expanded at a very high speed in recent decades, the decline in poverty could not keep pace with it. In fact in some regions of the poor countries of the world, the absolute poverty has actually accentuated. One of the reasons behind this sorry state of affairs is the fact that often we concentrate on the amelioration of the income poverty only and neglect the other dimension of it. In fact, poverty is not a uni-dimensional concept; rather it is a multidimensional one. It continued for many years in a large number of countries. In the developing countries poverty alleviation program is an important issue for sustainable development of the country. For improving the quality of life and well-being of the poor people we have to understand the problem of poor people and it is also necessary to know the relevant dimensions of poverty. Because of multi-dimensional nature, poverty is not easy to classify. Conceptual and methodological differences in defining poverty can lead to the identification of different individuals and groups as being poor. As a consequence, policy-makers are facing a problem of measurement as well as the definition of poverty.

Under the phase of economic reforms in the low income countries as well as throughout the world, the number of people living in chronic poverty has increased during the last five years. In this period the number of chronically poor people lies on a range between 320-443 million. The characteristics of chronic poor people not only depends on low income but also on deficiencies like hunger, malnutrition, illiteracy, lack of safe drinking water and primary health services, social discrimination, lack of physical security and political exploitation. Here poverty, environment and development have a complex relationship in the developing countries. Since 1991, growth rate is increasing from 5.4% to 8% and Below Poverty Line (BPL) has been reduced to 26.7% in India.

In the last century the global population increased from 1.6 to 6.1 billions. The massive population growth leads to slow down of per capita income growth due to resource constraint, and degrading environment due to urbanization, industrialization and expansion of agricultural land. After the reforms the poverty reduction policies are paying attention on two basic goals: (1) achieving macroeconomic stability to strengthen economic growth, and (2) providing the basic factors (health, education, water and sanitation) for human development.

In this paper we have chosen India as our case study as it has the dubious distinction of having the largest number of poor in the world in a single country for a long time in spite of adoption of poverty alleviation measures for more than three decades. It is also a fact that while Indian economists have contributed largely to sophistication of poverty measurement techniques yet officially it has relied all along on the crude head-count ratio for estimating the poor, a technique which cannot fulfil any of the axioms desirable for poverty index. India also recently has experienced the paradoxical situation of high rate of GDP growth consistently for the last five years along with a vast mass of poor people estimated approximately 30-40% by the national poverty line (which is lower than the international one dollar per day income level) and more than 47% by one dollar per day poverty norm in a country of one billion plus population. Further the acuteness in poverty in India is not fully appreciated even in the richer parts of the country, not to speak of donor countries of India mostly belonging to the rich North as more than 75% of Indian poverty is concentrated in 4-5 states in India. The impact of high rate of income growth is also not felt properly due to rising inequalities of income and wealth in this country during the last two decades or so as a consequence of economic reforms. All these features make Indian case study an interesting and important one for poverty research.

I.1 Four Approaches of Poverty

To define poverty, four approaches are emphasized. These are the monetary approach, the capabilities approach, the social exclusion approach and the participatory approach. In the monetary approach a poverty line is defined in terms of the income sufficient to achieve a minimum standard of living and to estimate the poverty line, both food and non-food expenditure is taken into account. This is most usually used to define and measure poverty. But this measure has a limitation, i.e. it does not cover the accessibility of social services which has a boosting power for a person's well-being.

The capability approach pioneered by Amartya Sen, highlights that income is the only important thing which increases the capabilities of individuals and thereby allows them to perform in a society. The ultimate objective of capabilities is the ability to lead a long life, to function without chronic morbidity, to be capable of reading, writing and performing numerical tasks and to be able to move from place to place. According to this approach a person who is not capable to achieve the minimum acceptable standard is poor. The requirement of income or resources to achieve the same capabilities may differ from person to person. It may be said that, a capability-poor person may not be income-poor. To measure the capabilities, in practice we use literacy, life expectancy and some other qualitative data. Capability approach covers the neglected part of the monetary approach.

When the individuals or groups are not capable to participate fully in their society that means they have a distance from income, capabilities and other daily necessities, then the concept of social exclusion comes out. According to this approach, poverty is a creation of social activity and may be divided in a group of characteristics, like women, the aged, handicapped, racial or ethnic categories, rather than as individuals. This approach not only focuses on the result of social exclusion but also on the processes that lead to it.

The participatory approach depends on the views of poor people themselves. They themselves decide the meaning of poor, and they determine the degree of poverty. This approach depends on the background method of analysis and which helps to understand the different dimensions of poverty within the social, cultural, economic and political environment of a locality.

According to the empirical evidences it is seen that, if we use the different approaches in a country in a particular period the rate of poverty will differ. As a consequence, the policy option can differ accordingly. Increasing money income is a solution of monetary approach but in the capability approach the social provision of goods to achieve health, nutrition and education are the main targets. The importance of increase in money income in the capability approach means to promote some capabilities. Growth can solve the problem of absolute poverty but not the relative poverty. Monetary approach and capability approach are actually dealing with the absolute poverty but the social exclusion is mainly concerned with the relative poverty.

I.2. Other Dimensions of Poverty

Besides the four dimensions of poverty as mentioned above we can trace the other dimensions also, some of which may be enumerated as given below.

Both spatial and temporal dimensions of poverty are important for India and other developing countries where poverty is not uniformly distributed and where the dynamics of poverty changes is neither continuous nor of a definite pattern. We can also consider the international dimension of poverty expressed through the national poverty line and per capita one dollar per day and two dollar per day poverty norms accepted internationally. In the context of the ongoing globalization process and the trend of convergence observed in many sectors under it, it is also interesting to enquire whether this convergence process is applicable for the poverty norm also, i.e. whether the national poverty level and one dollar per day poverty norm gradually converge with the two dollar per day norm as formulated and updated from time to time. In many countries like India another dimension of poverty is quite relevant especially in the rural areas and where the population largely depends for their standard of living on natural and forest resources. We may term this as an ecological dimension of poverty which results due to continuous degradation of environment due to unsustainable pattern of economic growth neglecting the environmental cost in this process.

II. The Nature of Chronic Poverty

The term chronic poverty is used to describe the extreme poverty which continues for 'a long time'—may be many years or an entire life, or even generation to generation. The people who are chronically poor, suffer from multiple dimensions of poverty which we have already discussed in the earlier section. Combinations of capability and accessibility deprivation, low levels of material assets, and sociopolitical marginality keep them poor over long periods. The people who are permanently deprived of well-being, also eventually come under the chronic group.

II.1 Causes of Chronic Poverty

This is very rare that the chronic poverty is depending on a single cause. All the causes may be sub-grouped in Economic, Social, Political and Environmental causes. This is shown by the following table.

TABLE 1

Causes of Chronic Poverty

Economic	Low productivity, Lack of skills, 'Poor' economic policies, Economic shocks, Terms of trade, Technological backwardness/lack of R&D, Globalization
Social	Discrimination (gender, age, ethnicity, caste, race, impairment), High fertility and dependency ratios, Poor health and HIV/AIDS, Inequality, Lack of trust/social capital, Culture of poverty
Political	Bad governance, Insecurity, Violent conflict, Domination by regional/global superpowers, Globalization
Environmental	Low quality natural resources, Environmental degradation, Disasters (flood, drought, earthquake, etc.), Remoteness and lack of access, Propensity for disease ('the Tropics')

Source: *The Chronic Poverty Report*, 2008-09.

II.2 Poverty in India

In the 21[s1] century poverty, environment and development have a complex relationship in the developing countries like India. As a developing country, India has some strategic plans regarding economic development, reducing poverty and to stop environmental degradation. In 1971, Dandekar and Rath first estimated the poverty levels in India and also gave a suggestion on poverty alleviation which is helpful for policy-making. From 1991 as an indicator of development, growth rate is increasing from 5.4% to 8% and Below Poverty Line reduced to 26.7% . According to the World Development Indicators 1999, it is seen that 35% of the population is poor in terms of its own national poverty level in 1994, but in terms of international level it was 47% in the same year. In the global perspective, in the last century population increased from 1.6 billions to 6.1 billions. The massive population growth leads to slow down of per capita income due to resource

TABLE 2

'Assets' and the Chronic Poor

Dimension of poverty/well-being	*Characteristics of the chronically poor*	*Significance for the chronically poor, in terms of ease/difficulty of realisation or upward mobility*
Physical assets: land, livestock, house, other	• Landless, near landless, marginal land; limited access to necessary inputs (e.g. labour, irrigation) • Physical assets few of poor quality, vulnerable to theft	• Few opportunities to accumulate assets • Redistribution policies are increasingly rare and are usually poorly implemented • High likelihood of losing physical assets
Financial assets and substitutes: income (trade, wages, rents, remittances, other); savings; investment; consumption	• High or low variability around a low mean, depending on wider socio-economic context • Few opportunities for diversification that would permit income or asset augmentation	• Microfinance has been extended to some poor people but there has not been a microfinance revolution • The poorest rarely access microfinance institutions
Geographical capital	• Remoteness, marginality. lack of physical and social infrastructure, poor environment.	• Opportunities for migration out of marginal areas depend on other forms of social and economic capital
Health and nutrition assets	• Vulnerability associated with disease, impairment, age • Ultra-poor consuming <80% of required calories but spending >80% on food	• Ill-health often has catastrophic impacts on other assets • HIV/AIDS reshaping health levels in many countries

Education and training assets	• Poor (or no) education • Few opportunities to develop new skills • Reliance on coping strategies	• Difficulty of maintaining enrolment (in terms of cost and time), especially for girls, especially up to secondary/technical level • Key inter-generational exit route, but highly dependent on labour market
Social and political assets	• Vulnerability associated with age, disability, gender, caste, ethnicity, religion • High levels of dependence and adverse incorporation into patron-client relations	• Multi-stranded patronage webs difficult to extricate oneself from, and are often passed on inter-generationally • Chronic poor generally have little or not voice in policy or governance
Security assets	• Vulnerability to violence, including domestic violence	• Very significant, but often not understood by researchers and policy-makers
Psychological assets	• Effects of long-term poverty and dependence on dignity; sense of self; risk aversion	• Labels (by public policy and by civil society) tend to stick over long periods and be transmitted inter-generationally

Source: Hulme, David, Karen Moore and Andrew Shepherd (2001).

TABLE 3

Achieving the Five Freedoms for the Chronically Poor: Implications for Public Policy

The five freedoms	*Significance for the CP*	*Difficulties in realization*
Political freedoms	Democracy has demonstrated effectiveness in preventing economic disasters and famines.	Role of organized opposition and extensive public debate is particularly important (e.g. about social opportunities).
Economic facilities	Market mechanism as a basic arrangement for mutually advantageous interactions for all.	Efficiency contributions do not guarantee distributional equity. Social opportunities to participate in the market may be constrained. Asymmetries of information and power realized through unregulated functioning of markets.
Social opportunities	Basic education, elementary medical facilities, key resource (e.g. land for agriculture) availability.	Can be significantly enhanced by public policy on health, education, land reform. Human development is basic, not a luxury for rich societies only.
Transparency guarantees	Crisis tends to be unequally shared. Transparency of information about government and business necessary to create broad relations of trust in society as basis of economic security.	Greater transparency and open public discussion of critical issues is helped by press freedom, media independence, expansion of basic education especially for women, and enhancement of individual economic independence through employment especially for women.
Protective security	Access to private and community insurance and safety nets likely to be weak for CP, so state-assured safety nets are very important.	Realizing access to safety nets requires effective governance and/or politically educated citizens.

Source: Adapted from Sen (2000).

TABLE 4

Head-count Ratio and Number of Poor Persons Below Poverty Line in India (Combined)

States/UTs	*Head-count Ratio*				*Number of Poor Persons (in Lakh)*			
	1973-74	*1983*	*1993-94*	*2004-05*	*1973-74*	*1983*	*1993-94*	*2004-05*
	1	*2*	*3*	*4*	*5*	*6*	*7*	*8*
Andhra Pradesh	48.9	28.9	22.2	15.8	225.7	164.6	154.0	126.1
Assam	51.2	40.5	40.9	19.7	81.8	77.7	96.4	55.8
Bihar	61.9	62.2	55.0	41.4	370.6	462.1	493.4	169.2
Delhi	49.6	26.2	14.7	14.7	22.8	18.4	15.5	22.9
Goa	44.3	18.9	14.9	13.8	4.2	2.2	1.9	2.0
Gujarat	48.2	32.8	24.2	16.8	132.4	117.9	105.2	90.7
Haryana	35.4	21.4	25.1	14.0	38.3	29.6	43.9	32.1
Himachal Pradesh	26.4	16.4	28.4	10.0	9.7	7.4	15.9	6.4
Karnataka	54.5	38.2	33.2	25.0	170.7	149.8	156.5	138,9
Kerala	59.4	40.4	25.4	15.0	135.5	106.4	76.4	49 6
Madhya Pradesh	61.8	49.8	42.5	38.0	276.3	274.0	298.5	249.7
Maharashtra	53.2	43.4	36.9	30.7	287.4	290.9	305 2	317.4
Orissa	66.2	65.3	48 6	46.4	154.5	181.3	160.6	178.5

(Contd.)

TABLE 4 (*Contd.*)

	1	*2*	*3*	*4*	*5*	*6*	*7*	*8*
Punjab	28 2	16.2	11.8	8.4	40.5	24.6	25.1	21.6
Rajasthan	46.1	34.5	27.4	22.1	128.5	126.4	128.5	134.9
Tamil Nadu	54.9	51.7	35.0	22.5	239.5	260.1	202.1	145.6
Uttar Pradesh	57.1	47.1	40.9	32.8	535.7	556.7	604.5	590.0
West Bengal	63.4	54.9	55.7	24.7	299.3	318.7	254.6	208.4
Chhattisgarh	—	—	—	40.9	—	—	—	91.0
Jharkhand	—	—	—	40.3	—	—	—	116.4
Uttarakhand	—	—	—	39.6	—	—	—	36.0
Chandigarh	28.0	23.8	11.4	7.1	0.8	1.2	0.8	0.7
Dadra & Nagar Haveli	46.6	15.7	50.8	33.2	0,4	0.2	0.8	0.8
J&K	40.8	24.2	25.2	5.4	20.5	15.6	20.9	5.9
All India	54.9	44.5	36.0	27.5	3213.4	3229.0	3203.7	3017.2

Notes:
1. Poverty Ratio of Assam is used for Sikkim, Arunachal Pradesh, Meghalaya, Mizoram. Manipur, Nagaland, and Tripura.
2. Poverty Ratio of Tamil Nadu is used for Pondicherry and Andaman and Nicobar Islands (A&N Islands).
3. Poverty Ratio of Kerala is used for Lakshadweep.
4. Urban Poverty Ratio of Punjab is used for both rural and urban poverty of Chandigarh.
5. Poverty Line of Maharashtra and expenditure distribution of Goa is used to estimate poverty ratio of Goa.
6. Poverty Line of Maharashtra and expenditure distribution of Dadra & Nagar Haveli is used to estimate poverty ratio of Dadra & Nagar Haveli.

Source: Planning Commission.

constraint, and degrading environment due to habitation, industrialization and expansion of agricultural land.

Poverty alleviation is one of the central objectives of India's social and economic development. In 1934 M. Visveswaraya stated in his *Planned Economy for India* that without the concept of planned economy for achieving the ultimate objective of India (poverty eradication) is not fulfilled. Under the leadership of Subhas Ch. Bose the Indian National Congress established the National Planning Committee under the chairmanship of Jawaharlal Nehru before the Second World War. In 1938 the Committee was constituted with some objectives by ensuring the optimum standard of living for the masses and eradication of poverty. At the same time Visveswaraya also told that for achieving the target, rapid growth through industrialization and import substitution with leading role of the state were needed for India's economic development. In 1950 (after independence), for preparing national developmental plans a Planning Commission was set-up. Some annual plans and eleven five-year plans have been formulated by the Commission. In 1974 Srinivasan and Bardhan stated that according to the Commission rapid growth generation is not the only factor which can eradicate the poverty but redistributive transfers among the poor are also necessary. Therefore, it may be stated that for poverty alleviation rapid growth of capital formation as well growth of income are the major instruments.

TABLE 5

Indian Poverty Rates Recorded by NSS

Year	*Round*	*Poverty Rate (%)*	*Poverty Reduction per year (%)*
1977-78	32	51.3	—
1983	38	44.5	1.3
1987-88	43	38.9	1.2
1993-94	50	36.0	0.5
1999-00	55	(26.09)	not comparable
2004-05	61	27.5	0.8

Source: Different NSS Reports.

According to Reuters report (10th August, 2007), 77 percent of

Indians, i.e. near about 836 million people live on less than half a dollar a day in one of the world's emerging economies. According to National Commission for Enterprises in the Unorganized Sector (NCEUS) most of those living on below 20 rupees (50 US cents) per day were from the informal labour sector with no job or social security, living in abject poverty. Around 26 percent of India's population lives below the poverty line, which defined as 12 rupees per day.

Since the early 1990s the economic liberalization has created a 300 million-strong middle class and led to an average annual economic growth of 8.6 percent over the last four years, but millions of the country's poor remain untouched by the boom. According to the report, based on data from 2004-05, 92 percent of India's total workforce of 457 million were employed as agricultural labourers and farmers, or in jobs such as working in quaries, brick kilns or as street vendors. The report said the majority of those working and living under "miserable conditions" were lower castes, tribal people and Muslims and the most disadvantaged of these were women, migrant workers and children. This is the other world which can be characterized as the India of the Common People, constituting more than three-fourths of the population and consisting of all those whom the growth has, by and large, bypassed.

II.3 Economic Growth and Poverty

Since 1950 the average growth rate for the period 1950-98 is near about 4% per year. After the end of three decades of planning this average was even less at 3.5% per year, which is well known as *Hindu rate of growth*. During the three decades (1950-81) due to massive failure to achieve the growth targets the percentage of poor population has fluctuated around 50%. During the 1980s growth rate increased to over 5% per year, and side by side the poverty rate also declined, but this unsustainable rapid growth of eighties was basically based on fiscal expansion which was financed by borrowing at home and abroad (a significant proportion of which was at commercial rates) and ended with the major consequence of macroeconomic crisis of 1991. During the economic reform era, after almost zero growth during the year 1991-92 of crisis, the growth was adjusted in 1992-93 and the GDP growth rate averaged 6.8% per year; in 1996-97 the GDP growth rate again declined to 5% but during 1998-99 it was again in good health, i.e 6.8% and in 1999-2000 it is estimated at 6.4%. With rapid growth rate, education and health sectors also received high priority in the developmental plans. The government attempts to provide free and compulsory primary education, and the current standard of living improved. The literacy rate also increased, but due to poverty side by side the number of drop-out school children was also increased because the poor parents are not able to support their school-going wards, who are actually encouraged to earn something. With these, some other schemes like the Public Distribution System (PDS) for food grains and other essential commodities, and various employment generation schemes have been adopted for alleviating poverty.

III. Case Studies of Ecological Poverty

The major part of the people of India lives in rural society and for their livelihood they depend on the common property resources. For the rural poor, improving the *gross natural product* is more important than increasing the *gross national product* (*Agarwal*, 1985). The challenge of the rural people is to build a base of natural capital which can support to build a healthy local economy. Rural villages have a perfect ecosystem with fine balance. These fine tuned ecosystems may be damaged or become fragile due to overuse. According to International Fund for Agricultural Development (IFAD) (1992) near about four billion people lived in 114 developing countries and out of that near 2.5 billion lived in rural areas and near about half of them are poor. It is seen that near about half of the rural population have no access to the safe drinking water, and even very few people have the accessibility of irrigation water for their agricultural activity. It is also found that the life expectancy of the people of the developing world is just 40 years (*IFAD*, 1992).

According to Global Poverty Report (*IFAD*, 2001) globally, 1.2 billion people live in 'extreme poverty': they subsist on less than one dollar a day. Seventy five per cent of the poor work and live in rural areas; 60% are expected to do so in 2020 and 50% in 2035. According to IFAD global rural poverty portal, if we consider the world statistics it is seen that more than one billion people in the world live on less than US $1 a day and 2.7 billion struggle to

Map of India Showing Locations of Case Studies

survive on less than US $2 per day. If we consider the number, it is seen that more than 800 million people go to bed hungry every day, including 300 million children. Every 3.6 seconds a person dies of starvation and most of those who die are children under age of 5; every year 6 million children die from malnutrition before their fifth birthday. Rural poverty of the developing countries is not the income poverty, and due to high dependence on common property resources it is basically ecological poverty.

Following are five case studies of rural villages in India that have succeeded to remove the ecological poverty by creating sustainable economic wealth. Most of the people in these areas have very small amount of land and they are poor. All the people depend on common property resources for their continued existence. These villages present an advantage for stimulating the economy, with establishing the cooperation among the villagers.

The first case is a sub-Himalayan range in north India, Sukhomajri; the second is Ralegan Siddhi, a village in the state of Maharashtra. The reform initiatives in these two villages are now more than 20 years old, offering a view of how natural assets are built over time. The third study, from the dry and hilly Alwar region of Rajasthan state, follows changes which began more than 12 years ago. In these three cases, the momentum for reform came from individuals and outside the government. The experiments received much attention. And the government learned some lessons from these success cases. Then the government took initiative from the case of Arabari forest, East Medinipore, West Bengal; this is the fourth case of our study. After that in 1996, the state of Madhya Pradesh initiated a statewide program for watershed development based on the model of Ralegan Siddhi. These case studies demonstrated that eco-restoration is possible in any situation and stimulating the local economy can alleviate the rural poverty which is basically ecological poverty and not the income poverty.

III.1 Watershed and Forest Management in Sukhomajri

Sukhomajri village is located near Chandigarh city, in the foot hills of Shiwalik in northwest India in the state of Haryana. In 1976 this village was thinly populated with only 455 inhabitants (59 families). At that time the facility of agricultural activity was very poor and the probability of soil erosion was very high. The village was homogeneous in terms of caste. According to P.R. Mishra, annual average rainfall was 1,137 mm. Out of 59 families 37 families have less than one hectare of land, 20 families owned one or two hectares of land and only two families have more land, but half of the total land of the village was panchayat land or community land. The agricultural land was very unproductive and every year some cultivated land was destroyed due to soil erosion. So, to supplement their income as well as to survive the villagers forced to keep a large number of livestock for selling milk. But at the same time villagers faced the problem of fodder scarcity and sometimes they had to import the fodder for their livestock from other villages.

In 1976, due to low rain fall, villagers realized the potentiality of dam or water reservoir. Daulat Ram, one of the enterprising villagers took the initiative to establish an earthen dam for irrigation. In 1978, the second dam was built with the help of Ford Foundation. A few farmers started to irrigate water-intensive crops like paddy and sugarcane, even though the project was supposed to provide only modest, supplemental irrigation. To get equal share of water, the villagers established a "Water Users" Association and each family had a representative. As crop production was increased, people sold their excess livestock and it was seen that goat population decreased from 206 in 1977 to 32 in 1983.

The villagers also developed grass land in the forest area and built a natural resource base of the village. To rely on the initiative of the villagers, the government gave the rights to Sukhomajri and a nearby forest village Dhamala for harvesting bhabbar grass. The villagers also harvested khar trees. In 1997, the khar trees could generate proceeds of about Rs. 30 million. Following table gives the Progress in Sukhomajri at a glance.

III.2 Initiative of Anna Hazare in Ralegan

Ralegan Siddhi is situated in Parner Taluka of Ahmednagar district. The Ahmednagar district lies in the scarce agro-climatic zone. Ahmednagar district has an area of 16762 sq. km. This district falls in the drought prone area of Maharashtra state, receiving 500 to 700 mm of annual rainfall with long dry spells. The rainfall is unevenly distributed during the monsoon period. The soils of the district are alkaline. The total geographical area

TABLE 6

Progress in Sukhomajri

Particulars	*Benefits*
Crops	The yield rates of wheat and maize, the two main staples, increased by more than 50 per cent between 1977 and 1986.
Grass	Grass production rose 75-fold, from 40 kg per hectare in 1976 to 3 metric tons per hectare in 1992.
Milk	With more fodder available from the forest, villagers have shifted their livestock from goats to buffalo. The number of goats dropped from 246 to 10 during 1977 to 1986, while the number of buffaloes rose from 79 to 291. As a result, daily milk production rose from 334 liters to 579 liters.
Trees	In the watershed, the number of trees increased almost 100 times, from 13 per hectare to 1,292 per hectare, between 1976 and 1992.

Source: Agarwal, Anil and Sunita Narain (2000).

TABLE 7

Changes in Production in Sukhomajri Village (1977-86)

Commodity	*1977*	*1986*	*Measure*
Wheat	40.60	63.60	Tonnes
	0.68	1.43	Tonnes/ha
Maize	40.90	54.30	Tonnes
	0.61	1.22	Tonnes/ha
Grass	0.04	3 (in 1992)	Tonnes/ha
Goats	246 (in 1972)	10	Number
Buffaloes	79 (in 1972)	291	Number
Milk	334	579	Litters/Day
Tree Density	13	1292	Number/ha

Source: povertyenvironment.net/files/CASE%20India.pdf

of the district is 17.02 lakh ha out of which 13.16 lakh ha is cultivable land. Near about 10% of the cultivable land is irrigated by the different canals and wells. Sugarcane is the main cash crop of the district and the principle grain crops are bajra in kharif and jawar and wheat.

In 1975, when Anna Hazare, a retired army man, found the village reeling under drought, poverty, debt, and unemployment,

Fig. 1: Location Map of Case Study Village

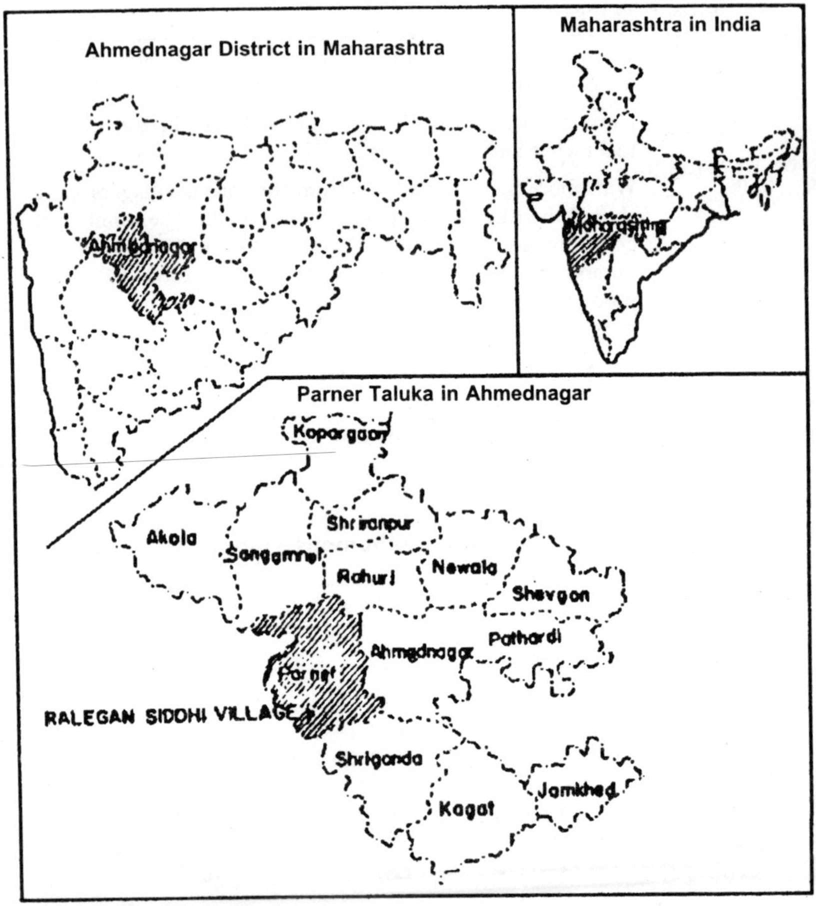

Source: Sharma, Prem N. and Mohan P. Wagley (1996).

he decided to change the economic structure of the village with the collective support of all the villagers. The first program taken up by Hazare was that of environmental regeneration. Rain water harvesting was undertaken through watershed development, i.e. by percolation tank, contour bunding, construction of check-dams and large scale tree plantation (the village planted 300,000 to 400,000 trees in and around the village, using a government forestry program). These programs revitalised the existing wells in the village. Many new wells were dug and utilization of water was regulated through six co-operative irrigation societies. As a result, about 700 acres of land were brought under irrigation. In addition, another 500 acres were brought under irrigation by a lift irrigation scheme from the Kukdi canal.

TABLE 8

Progress in Ralegan Village during 1975-76 to 1985-86

Commodity	*1975-76*	*1985-86*	*Measure*
Bajra	20	150	Hectare
Jowar	50	250	Hectare
Vegetables	2	60	Hectare
Wheat	1	23	Hectare
Oil Seed	Nil	17	Hectare

Source: http://www.umiacs.umd.edu/-~venu/ANNA/economic.html

As a result of the above changes total agricultural production increased from 294.3 tonnes in 1975-76 to 1386.2 tonnes in 1985-86. At current prices, it means an increase from Rs. 3.46 lakh to Rs. 31.73 lakh, i.e. 4.7 fold increases in quantity and a 9-fold increase in value.

Milk production was the secondary occupation in Ralegan. The milk production was also increased by replacing the local cows with the crossbred ones, which gave a high milk yield. The number of milch cattle has also been growing. In 1987 there were 574 milch cattle (including 65 cross-bred cows) as compared to 255 in 1981. Around 350 litres of milk were collected per day and sent to a co-operative dairy in Ahmednagar.

With the initiative of Anna Hazare the village also runs a grain bank. The grain bank also helps in preventing distress sale of grain at a low price at the time of harvest and purchase at a higher price from the market during the lean period before the next crop. In 1975-76 only 20 quintals of foodgrain were collected during the year whereas this collection increased to 40 quintals in 1985-86. Some other secondary occupations like poultry farming were also encouraged, but with limited success. Out of the 10 persons who received training only one person was actually running poultry successfully. Living standard was also increasing in Ralegan through improved educational facility. It is seen that in 1975-76 only 20 persons were in job with an average earning of Rs. 200 per month whereas in 1985-86 over 90 persons were earning an average of Rs. 1000 p.m. in their city jobs.

III.3 Initiative of Tarun Bharat Sangha (TBS) in Gopalpura, Rajasthan

The Gopalpura villages situated in the Aravalli hills of Rajasthan's Alwar district is a poor and drought prone village. The region gets roughly 600 mm of rainfall a year. Surface water evaporates quickly due to high temperature in this region. Agricultural activity was not so impressive and industrial activity was also very few and all are located adjacent to the city. Therefore, rate of migration was very high. People were struggling to survive in this region.

In 1986, under the leadership of a local NGO named as Tarun Bharat Sangha, the people of Gopalpura village built three earthen structures for water harvesting on their fields and grazing lands to collect monsoon rains for irrigate their fields and to recharge the ground water. These structures are called *Johads*. With the help of Tarun Bharat Sangha over the last 15 years, almost 2,500 water conservation structures were built in 500 villages which begun to revive the local tradition of water harvesting. This recharging of ground water has led to sustainable agricultural development which has allowed local farmers to survive consecutive drought-years and migration has been greatly arrested. Studies show that the village domestic product has increased in proportion to the investments made in water conservation. TBS supplied some materials and equipments, such as cement and diesel for tractors. Villagers were required to contribute labour and other materials. The total investment came to Rs. 15 crore. Despite their extreme poverty, villagers contributed an amazing 74% of the total, in cash or in kind. In each settlement, the village assembly met to plan for *johad*. For fishing rights villagers had a tussle with the government, and lastly the villagers have won the rights.

III.4 Arabari Forest Range, Medinipore Forest Division, West Bengal

The Arabari forest is situated in the Eastern part of the West Medinipore district of West Bengal in Eastern India. The forest range is spread throughout the two police stations of Salboni and Keshpur. However, the forest is under the West Medinipore political boundary but it is under the jurisdiction of East Medinipore Divisional Forest Office. The area under the Keshpur police station is 186 sq. km. comprising with 11 mouzas. The caste structure of this area is heterogeneous, viz., the Santals, low castes and some general castes. The quality of forest in Medinipur is densely Sal, but severe forest degradation is noted. The forest fringers are basically small farmers and daily labourers. Still now they use the conventional or traditional agricultural procedure and due to increase in population density, conversion of forest land to agriculture land is increasing. Arabari range comprises 11 villages viz., (1) Chandmura, (2) Sat Sole, (3) Sakhi Sole, (4) Mahis Dubi, (5) Majhi Dubi, (6) Sapdiha, (7) Urami, (8) Ghuchi Sole, (9) Ankra Sole, (10) Sank Sole, (11) Dhobhani. In 1972, an innovative plan to contain the deforestation problem was launched on an experimental basis in the forest-fringe villages of Arabari. It involved local villagers in protecting coppices of Sal trees in return for free usufruct rights on all non-timber forest products, additional employment, and a promise of 25% share of the net cash benefits from the sale of short rotation Sal poles. About 1,270 hectares of degraded Sal forests were taken up for revival on a pilot basis. Initially, 618 families, comprising a population of 3,607, were involved through "forest protection committees". Sal and its associates in forests yield many non-timber forest products like Sal leaves and seeds, mushrooms, tasar silk cocoons, medicinal plants, edible roots and tubers, etc. which motivates the poor villagers in protecting the coppices during their gestation period.

By mid-1970s, the West Bengal Government realized that, if the people's needs were ignored, it would be impossible to save the forests. The National Commission on Agriculture (1973) recommended, among other things, "social forestry" on land unsuitable for agriculture to relieve pressure on forests and to help in soil and water conservation. The Forest Conservation Act of 1988 also helped to prevent uncontrolled conversion of forest land for non-forestry purposes and made compensatory afforestation obligatory in such cases. The success of social forestry programme on non-forest land has led the planners to realize that it is a viable land use system and an important tool in development of rural areas where large scale employment is possible through it.

Encouraged by the experience of the Arabari experiment, the State Government decided in 1987 to encourage forest-fringe population to actively participate in managing and rehabilitating degraded forests all over south-west Bengal. This movement spread like a wild fire. Though informal and voluntary at first, it acquired the character of a formal institution when, in 1990, the State Government officially recognized the forest protection committees (FPC) in south-west Bengal.[1]

Fig. 2: Map of West Medinipur District (Showing the different Blocks)

PASCHIM MEDINIPUR

Source: Saha, S., 2003.

Location of Arabari Range

Fig. 3: Detailed Map of Arabari Range

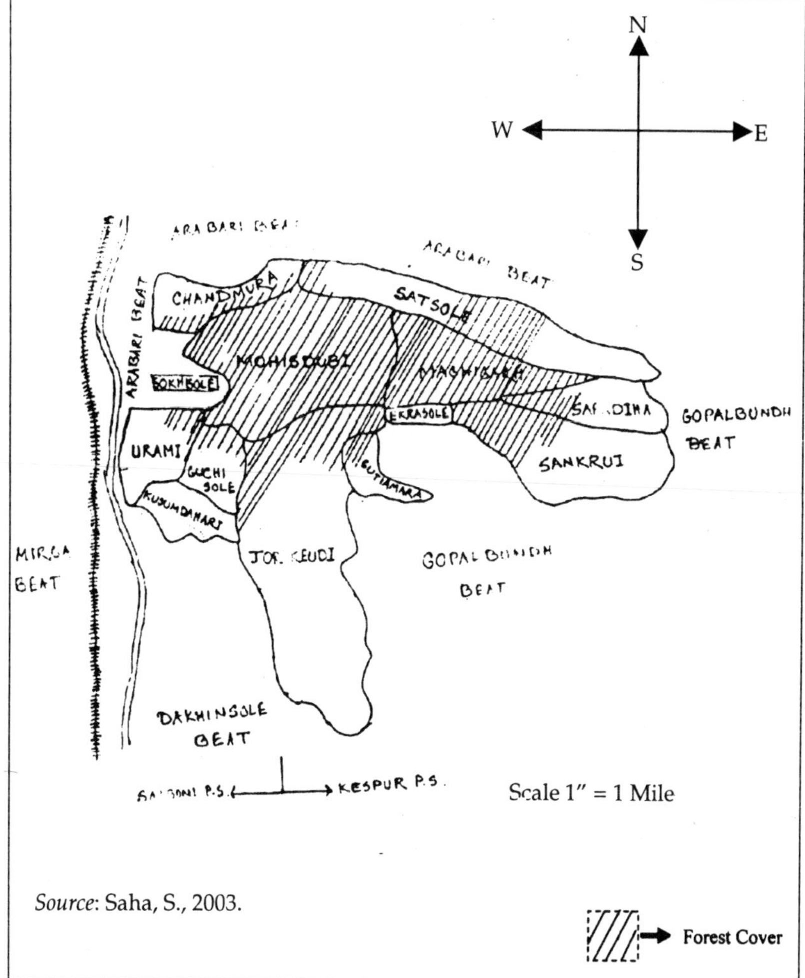

Source: Saha, S., 2003.

III.5 Jhabua, Madhya Pradesh

Jhabua is a predominantly tribal district located in the western part of Madhya Pradesh. It is surrounded by Panchamahal and Baroda districts of Gujarat, Banswara district of Rajasthan and Dhar and Ratlam districts of Madhya Pradesh. River Narmada forms the southern boundary of the district. The total area is 6793 sq. kms. According to 2001 census there are about 1313 inhabited villages. About 85% of population is tribal while 3% population belongs to Schedule Castes. 47 per cent of the people live below the poverty line. Most part of Jhabua is without any forest cover because of low fertility of land and soil erosionteven mere existence becomes a hard problem and with the failure of rains the Bhils take to crime.

After getting the inspiration form Anna Hazare, with the help of government, Digvijay Singh took the initiative to change the economic condition of Jhabua district through rainwater harvesting. In 1996, Rajiv Gandhi Mission on Watershed Development was started. Presently 249 micro watersheds have been taken up under the schemes of EAS, DPAP, IWDP and IJRY. Total of 1,47,066 hectares has been taken under integrated watershed development with an estimated project cost of Rs. 5277 lakhs. In order to effectively launch the project, 12 governmental and 8 NGO Project Implementing Agencies (PIA) were carefully selected. Using RRA/PRA techniques and thematic maps from remote-sensing, action plans were prepared by P.I.A's for every micro watershed. Fodder development has been promoted on private as well as government land. Grass beds, silvi-pasture and pasture development has been given emphasis. 2950 hectares of land has been taken up for fodder production.

Assisting and ensuring natural regeneration from dormant seeds, coppices and gap planting has been taken up in 2900 ha. of degraded forest areas. Informal groups or village forest committees under JFM have been protecting these areas for last 2 to 3 years. Approximately 27.12 lakh plants are found to be regenerating in watershed areas. 350 community assisted water harvesting tanks have been constructed. All these tank sites were selected by the villagers themselves. Villagers have contributed 50% of the total costs. Besides other water conservation and harvesting measures, 83 percolation tanks and 21 sub-surface dykes have been constructed to artificially recharge the groundwater.

IV. Conclusion

All these show that it is necessary to understand the character of poverty if we want to alleviate it effectively. In all the five success stories mentioned above it has been observed that poverty could be reduced by ecological upgradation when conventional economic development could not help the poor. One of the main reasons in this context is the neglect of the value of environment in a country dominated by the poor as it is correctly regarded that poverty is the greatest polluter. However these shining examples of decentralized and eco-friendly development show that there are effective ways to control the chronic poverty by empowering the poor, rethinking the development initiative, focusing on community-based projects and giving women a voice. Of course in the current circumstances these targets are difficult to achieve due to existing vested interest who are strong, powerful and vocal minority and are the beneficiaries of the current unsustainable development paradigm. It is high time that the ecological links with the development process are recognized to renew the development process in a sustainable manner. Chronic poverty is not something that will remain chronic if we can successfully adopt an ecological approach to remove poverty. This is particularly true for India which is not only one of the 12 mega biodiversity regions of the world but also which is the habitat of most of the poor people of the world who depend on natural resources and the surrounding ecology for their existence and survival.

Note and Reference

1. hftp://www.pccdindia.com/pcoplescomm/forest _4a.htm.

References

Agarwal, Anil and Sunita Narain (2000): Redressing Ecological Poverty through Participatory Democracy: Case Studies from India, *PERI Working Paper No. DPE-00-01*, University of Massachusetts, Amherst.

Agarwal, Anil (1985): "Politics of the Environment," *State of India's Environment, 1984-85: The Second Citizens' Report*, New Delhi: Centre for Science and Environment.

Agarwal, Anil and Sunita Narain (1989): *Towards Green Villages: A Strategy for Environmentally Sound and Participatory Rural Development*, New Delhi: Centre for Science and Environment.

Chronic Poverty in India: Policy Responses (2007): Policy Brief, No. 2, Chronic Poverty Research Centre (CPRC), Institute for Development Policy and Management (IDPM), University of Manchester, Manchester, UK.

Hickey Sam and Sarah Bracking (2005): Exploring the Politics of Chronic Poverty: From Representation to a Politics of Justice? *World Development*, Vol. 33, No. 6, pp. 851-65.

Hulme, David, Karen Moore and Andrew Shepherd (2001): Chronic poverty: meanings and analytical frameworks, *CPRC Working Paper 2*, Chronic Poverty Research Centre, UK.

Mahapatra, Richard (1998): "Sukhomajri: Development Model." *State of India's Environment 1999: Citizens' Fifth Report*, edited by Anil Agarwal, Sunita Narain, and Srabani Sen, New Delhi, Centre for Science and Environment.

Mehta, Aasha Kapur and Amita Shah (2001): *Chronic Poverty in India: Overview Study*, CPRC Working Paper 7, Chronic Poverty Research Centre, Institute for Development Policy and Management (IDPM), University of Manchester, Manchester, UK.

Radhakrishna, R., K. Hanumantha Rao, C. Ravi and B. Sambi Reddy (2006): *Estimation and Determinants of Chronic Poverty in India: An Alternative Approach*, Working Paper 2006-07, IGIDR, India.

Rapid Poverty Reduction (2008): *Eleventh Five Year Plan*, Planning Commission, India.

Saha, S. (2003): *Role of Women in Joint Forest Management at Arabari Forest*, Project Report, Forest Research Institute, Dehradun.

Scott Lucy (2006), *Chronic Poverty and the Environment: A Vulnerability Perspective*, CPRC Working Paper 62, Chronic Poverty Research Centre, UK.

Sen, A. (1999): *Development as Freedom*. Oxford: Oxford University Press.

Sen, Raj K. and S. Mazra (2008): 'Fragile States and People: A Multidisciplinary Approach to the Concept and Measurement,' *Journal of Human Development*. Serial Publications, New Delhi.

Sharma, Prem N. and Mohan P. Wagley (1996): *Case Studies of People's Participation in Watershed Management in Asia*, Part I: Nepal, China and India, FAO. Kathmandu, Nepal.

The Chronic Poverty Report, 2008-09: Escaping Poverty Traps (2008): The Chronic Poverty Research Centre (CPRC), UK.

Web References

www.jhabua.nic.in
www.povertyenvironment.net
www.umiacs.umd.edu
www.pcedindia.com
www.tarunbharatsangh.org
www.rainwaterharvesting.org

SECTION IV

SUSTAINABLE DEVELOPMENT: SECTORAL ISSUES

Mining Sector in India under Globalization and a Sustainable Development Approach in Terms of EIA and CSR

Binayak Rath

Consequent upon the rise in demand of basic natural resources as well as their prices in early part of this decade, there has been a significant rise in flow of FDI to those sectors and their up-stream industries by the Multi National Corporations (MNCs). Hence, over past few years there has been mushroom growth of mining activities in different mining bearing areas of our country along with the growth of mineral processing industries, which have resulted in creating environmental damages and pollution of land, water and air and thereby adversely affecting the socio economic conditions of the people living in its surrounding areas. Moreover, since most of the minerals are located beneath the forest areas, the extraction of minerals leads to loss of forest coverage and its flora and fauna. In order to contain these issues and negative environmental impacts, it is advocated that the proponents of projects should undertake appropriate abatement measures to achieve sustainable development of the economy. Our earlier studies have established that mining companies and private minors have brought out a significant change in the land-use including loss of crop fields, irrigation opportunities, drainage system etc. and thereby affecting the social, economic and environmental dimensions of the surrounding areas, further affecting the whole ecosystem. It also leads to several diseases among the people. In order to contain these adverse impacts, many new initiatives have been taken by the Government of India vide new policies, regulations and guidelines. Some of the important policies are New Environmental Policy, 2004, the R&R policy, 2003 and the EIA Rules of 2006, under which the miners and the industries have to adhere to many strict standards including corporate social responsibility (CSR). In this paper we have attempted to examine the scope of EIA and CSR in attaining the goal of sustainable development by undertaking a case study in some of our back ward regions. It is argued in this paper that there are many new opportunities for the mining sectors to grow under the umbrella of CSR for which one has to adhere to a stakeholders' approach, of course, with a deep commitment from the top management/owners.

I Introduction

In view of our new industrial and trade policies in the 1990s, not only there is a significant flow of foreign direct investment to different sectors of our economy, but also many of Multi National Corporations (MNCs) have come forward to invest in those industries which are heavily dependent on natural resources like coal, iron ore, bauxite, manganese, limestone, etc. for which we have a natural advantage in comparison to other countries. In addition, consequent upon the rise in demand, the world prices of some of these basic minerals had gone up significantly to make some of the unprofitable mines to become profitable. For instance, as the iron ore price had gone up by 3 to 4 times during fast ten years, many miners had come forward to extract iron ore from once deserted mines. Thus, past few years have seen a mushroom growth of mining activities in different mining bearing areas of our country along with the growth of mineral processing industries. But all these developments are no unmixed blessing for our economy as most of the mining activities are creating environmental damages and pollution of land, water and air and thereby adversely affecting the socio-economic conditions of the people living in its surrounding areas. Moreover, since most of the minerals are located beneath the forest areas, the extraction of minerals leads to loss of forest coverage. The flora and fauna also deplete due to mining activities. It is noted that if such negative impacts are not remedied through some abatement measures, it would lead to unsustainable development of the economy.

On the other hand, the policies of the Government of India as well as that of the State Governments to grant subsidies for the development of the backward regions (i.e., the planned goal of regional development) have provided incentives for the miners and processing industries to go for more and more mining. As most of the companies find it profitable to establish their operations in these areas, several such areas have been urbanized in the past 10-15 years. This trend of urbanization in its turn has led to obvious advantages to the people of these areas, with the increase in number of facilities and employments for the local people. However, at the same time, this has led to several drawbacks for the locality and the local people. Establishment of the mining companies and private minors have brought out a significant change in the land-use including loss of crop fields, irrigation opportunities, drainage system, etc. and thereby affecting the social, economic and environmental dimensions of the surrounding areas, further affecting the whole eco-system. For example, the establishment of the new industries have led to deforestation and also increased the pollution levels, which in turn have had an adverse impact on the crops and vegetations including all flora and fauna. It also leads to several diseases among the people. In order to contain these adverse impacts, the MOEF, and other Ministries of the Government of India have devised many new policies, regulations and guidelines. The New Environmental Policy, 2004, the R&R policy, 2003 and the EIA Rules of 2006 are some of the recent landmark policies, under which the miners and the industries have to adhere to many strict standards. In this paper an attempt has been made to examine the role of EIA and CSR in execution of mining activities in some of our backward regions to achieve the goal of sustainable development.

II. Mining and Mineral Scenario of India

The mining and the mineral sector is not only the major contributor of material and economic health of the industrialized world but also it brings higher economic growth of many developing countries those are endowed with natural resources.

India, being endowed with this gift of nature, possesses as many as 84 minerals those are under active exploitation; comprising 4 fuels, 11 metallic, 49 non-metallic industrial and 20 minor minerals. A country wise comparison of mineral production *vis-a-vis* India's production shows that India has major advantages in a few of the basic minerals. For instance, its share is 2nd in Chromite, Barytes, Talc/Steatite and Pyrophyllite; 3rd in Coal and Lignite, 4th in Iron Ore and Kyanite/Silliminite/Andalusite; 6th in Bauxite and 7th in Manganese ore.

In view of these potentials and also owing to rapid growth of mineral processing and mineral-based industries, the growth of mineral industry in India has been phenomenal. It has grown just from Rs. 58.0 crores industry at the time of independence to a staggering Rs. 80,000 crores by 2006. As a result of this growth the share of mining industry in GDP has gradually increased. It is noted that during 1993-94 when its share was 2.57% it had grown to 2.64% in the year 2005-06. Moreover, It contributes significantly to the revenues of the Central and State Government through payments of royalty, forest compensation fees, dead rent, cess, sales tax, excise duties and custom duties. As most of the mines are located in many backward and tribal areas, it is argued that promotion of mining could be an important engine for regional development and employment. Besides in view of its contribution to export (i.e., 16% in terms of value), the mineral sector has become an important component of India's foreign trade.

Move Towards High Growth in Post-liberalisation Period

Alongside the economic liberalization introduced by GOI in 1991, a comprehensive National Mineral Policy (NMP) was announced in March 1993. The Mineral Policy opened the gates of Indian mineral industry to domestic and foreign investment, much of which was earlier reserved especially for the public sector. As a result of this new policy, many mining activities have flourished over the years. Particularly, the iron ore sector has made a rapid stride during the post liberalization period. A boom in the construction industry, accelerated by the 2008 Olympic Games in Beijing, pushed the international price of iron ore from $17 per tonne in 2000-01 to the present $55 per tonne. At its peak-in 2004-05—the price touched a mind-boggling $75 per tonne. The higher international prices have induced export of iron ore and approximately 50% of iron ore produced is being exported. Its additional production further contributes to regional redistribution and employment generation. There has been significant investment in the mining and mineral sector after liberalisation owing to flow of capital from public and private sector including foreign capital. Many new investors/companies have come forward to extract minerals from remote areas in the country. One such hot spots of exploration of minerals is the Bellary region in Karnataka. Many owners of once abandoned mining lease areas with relatively low grades of minerals and ore in Goa, Chhattisgarh and Maharashtra have applied to renew their lease and the existing ones have applied for capacity expansion. However, these developments have given rise to many socio-political and environmental problems.

With regard to the expansion of mining and mineral processing industries in our country and more particularly, its expansion in the remote areas there are two strands of arguments. While the activists and environmentalists argue that the local people will bear significant burden due to loss of private land, common property resources and also due to environmental degradations, which would lead to deterioration in the standard of living of native people; the cornucopias and develop mentalists see a number of gains for the local people with the help of proper rehabilitation and resettlement (R&R) as well as through CSR activities as a part of their social responsibility. They argue that the companies should behave ethically and should contribute to the cause of the nation. In view of these arguments Government of India has framed a new R&R policy and formed a policy called Corporate Social Responsibility (CSR) under which the companies are supposed to spend a portion (around 2%) of their turnover for the welfare of the local people and the surroundings. In spite of these developments and proactive measures of the Government, the promoters and developers are besieged with many new challenges.

Challenges Faced by the Promoters

In spite of favourable policy of the Government towards M&M sector and liberal flow of capital, the promoters are facing many challenges in terms of:

(a) Getting mining lease from the appropriate authority,
(b) Getting statutory clearances from government machinery like DGMS, SPCB, FD, and MOEF,
(c) To get local support to carry out mining activity in terms of implementation of satisfactory R&R,
(d) To develop a sustainable development criteria *vis-a-vis* loss of livelihood of the local people because of loss of private land as well common property resources (CPR),
(e) To develop a comprehensive methodology for EIA/SIA, which have become the statutory requirement, and
(f) To develop a framework for sustainability indicators as a tool for performance assessment/monitoring.

III: Impact of Mining and Mineral Processing on Environment

As most of the mining activities involve acquisition of both public and private land including a large portion of CPR, both the physical and socio-economic environment are going to be adversely affected. It is envisaged that most cumulative environmental destructions are caused due to mining activities in most of the countries. A number of issues, including loss of biodiversity, forest destruction, industrialization, loss of livelihood, pollution of land, water and air, urbanization and population growth have drawn the attention of the policy-makers and project developers/ proponents. The major environmental impacts identified are:

Physical Impacts

The physical impacts of any mining project are generally gauged in terms of its negative as well as positive effects on Land,

Water, and Air, which in turn impacts the Flora and Fauna. The land area is affected due to deforestation, soil erosion, soil degradation (as shown in the photograph 1) accumulation of solid and hazardous waste on land, atmospheric/climate changes and loss of bio diversity. Loss of forest also entails a severe burden on animals and birds as well as on the tribal community and low income groups. Mining also affects the availability of fresh water in the surrounding areas as mining activity may draw a lot of water from the surrounding region, or it may pollute water sources of the existing water bodies including village ponds, rivers and other small streams. Many a times the mines also intercept the ground water tables, so as to contaminate it from the source. Similarly, due to loss of bio diversity, blasting operations and movement of heavy vehicles in the area, the air environment gets polluted. The mineral processing industries also emit a lot of smoke to pollute the environment and to add to climate change and global warming (as exhibited in photograph 2). To quote from a study undertaken by NEERI, "the environment of Bellary Hospet region is highly polluted due to dust pollution and water pollution and anthropogenic exploitation of the forest wealth".

Moreover, as the shelter belts for birds and animals get destroyed they will search for new resettlement sites. Unless protection measures are undertaken by the promoters, the flora and fauna may face extinction. Hence, in order to find out some remedial measures to contain these negative impacts and to attain the goal of sustainable development, there has been a need to study the environmental effects in terms of the tool of environmental impact assessment (EIA). In undertaking such an exercise, the stakeholders of M&M activity have to play a lead role so that the impacts are estimated properly.

Socio-Economic Impacts

Mining and mineral processing activities not only affect the physical environment, but also influence the socio-economic, cultural and aesthetic values of the people living in the core as well as periphery zone of the mines and processing industries. These activities will bring significant changes in the level of production, consumption, employment generation in the area directly and indirectly and that may lead either to higher economic growth or decelerate the growth process if the remedial

Photograph 1: Showing the level of Deforestation and Soil Erosion in the Bellary Region of Karnataka.

Photograph 2: Showing the Level of Air Pollution Owing to Mineral Processing Industrial Activity in the Bellary Region of Karnataka.

measures are not undertaken with a human face. Even the resource use patterns would change so as to bring socio-economic changes in the region. Those socio-economic impacts and changes can be identifies as follows:

Positive Socio-economic Impacts of M&M

(i) Since most of the mining and mineral processing activities are labour-intensive, it will create additional employment opportunities for the local people directly and indirectly. Further the processing activities in the upstream and downstream industries, tertiary activities and associated growth of informal sector is likely push up employment. The studies undertaken by us in some of the mining site and downstream industries have established that there shall be higher employment opportunities for the local people.

(ii) As income levels of the people are expected to rise, it will lead to improve level of consumption and quality of lives of the people. Of course, there may be some people who may be adversely affected due to rise in demands of various goods and services and the consequent price rise.

(iii) It is expected that there will be increase in the commercial, business and shopping centers and subsequent urbanisation due to influx of population in the region to cater the needs of existing population as well as the immigrants.

(iv) The project proponents are likely to develop the social infrastructures like road, education, health care, etc. in the core and buffer zone and the benefits from those facilities will flow to the local people.

(v) In view of promotion of commercial and social infrastructure, the land value in the area is likely to rise whose benefit will flow to the local residents.

(vi) The processing industries will not only encourage development of auxiliary and ancillary units in the periphery, but will promote better use of the local resources whose inducement impacts will flow to the neighbourhood. Photographs 3 and 4 show how the Jindal steel plant near Bellary have converted the barren unproductive land into productive use.

(vii) The plantation and water harvesting structures along with proper training facilities undertaken by the developers of mining projects under CSR would also add to higher production of crops and the local people may adopt commercial crop production to reap the benefits.

(viii) Ancillary agricultural activities encouraged in the buffer zone will create additional employment and income generating opportunities.

(ix) Adds to revenues of the Central and State Governments.

(x) Above all, promotion of education and training under CSR will create new opportunities for the local people and also create new awareness.

Adverse Socio-economic Impacts of M&M Sector

Notwithstanding the above mentioned positive impacts of mining and mineral processing industries, there are many adverse impacts on the surrounding areas as a result of which the people in the local area bear the burden on these economic activities. As has been observed that one of the consequences of our development projects including mining and its associated activities has been the *loss of cultivable land in the rural areas and more so a great amount of forest land and revenue land*, which are the sources of living for a large number of deprived section of our population. Hence, all those persons dependent on land and other natural resources need to be resettled and rehabilitated, which is the greatest challenge for the project developers. Various studies have established that the loss of land coupled with loss access to other natural resources and owing to pollution of land, water and air have accelerated the existing poverty and intensification of the existing miseries of the rural poor. The adverse socio-economic impacts identified are:

(a) Loss of private and public land affects agricultural production in the surrounding area and the same would lead to loss of consumption of the project affected persons unless they are properly rehabilitated.

(b) Loss of land and CPR adversely affects the animal wealth (cows, buffaloes, goats and sheep) of the area and this leads to shortage milk, cow dung, etc.

(c) Dust deposition on crops fields accounts for poor crop

Photograph 3: Showing the Positive Impacts of Land Development by Jindal Steel Plant at Taranagar in the Bellary Region of Karnataka.

Photograph 4: Showing the Positive Impacts of Land Development and Infrastructure by Jindal Steel Plant at Taranagar in the Bellary Region of Karnataka.

productivity and dust deposition on cash crops like mango, tamarind and other affects their productivity.

(d) As loss of production and productivity impacts supply and demand of goods and services in the area due to higher employment opportunities, the price of the basic goods would rise and this will entail higher burden on the poor people.

(e) Dust in air and water pollution coupled with poor health and hygiene facilities in the core as well as buffer zone often result in many diseases.

(f) Influx of population due to migration of labour to the area from outside may lead to strain on the basic infrastructural facilities, such as, water, road, solid waste disposal, sewage, etc.

(g) It is noted by us during our visit to few project sites that most of the migrant labour, owing to shortage of dwellings, construct unauthorized structure that give rise to slums in the area. Due to formation of slums, people would be forced to live in unhealthy dwellings and surroundings.

(h) Blasting during mining operations often lead to land vibrations and further shocks from blasting are likely to disturb the building and structures of the local residents.

(i) If mining activities destroy some religious and social institutions, its revival or reconstruction at a different location sometimes lead to social conflicts.

(j) Many a times illegal mining activity would lead to accidents and also poses occupational hazards for the local people.

In view of these negative impacts, the quality of life of the local people in the surrounding area is likely to decline unless proper rehabilitation and resettlement measures are undertaken by the promoters of mining and processing industries, where the stakeholders have to play a lead role. Before investigating the scope of such intervention, one should examine the broad issues and challenges associated with R&R. By undertaking a critical analysis of R&R policies of many industrial projects, we have identified the following broad issues associated with rehabilitation and resettlement.

- Involuntary displacement is generally associated with land acquisition by the state for different development projects in the name of "Public Purposes". The definition of "public purpose" is itself a controversial issue in R&R.
- The valuation of land and the norms used in payment of compensation as well as the forms of compensation payment raise a number of controversies.
- When peoples' access to common property resource (CPR) for their living like grazing of animals, (as shown in Photograph 5) becomes limited, the sources of income generation for a number of groups of population is reduced. All those dependants should also be rehabilitated.
- Since involuntary displacement imposes major economic and social risks upon the project affected people and groups, how to reduce or avoid project-induced social disruption and impoverishment?
- How to plan and execute the rehabilitation and resettlement schemes to mitigate the socio-economic losses arising out of involuntary displacement?
- How best the Remedial Action Plans (RAPs) be implemented and monitored?
- How to identify the project affected persons (PAPs) or who is a PAP?
- How to resolve the problems when resettlement leads to conflict with the host community where settlement colonies are set-up by the Government?
- What should be the role of stakeholders such as the Government, bureaucracy, jurists, economists, anthropologists, sociologists, NGOs/activists and above all, that of the PAPs in tackling the problems of R&R?
- What should be the role of different actors in the process of resolution of social evils including political structure, bureaucracy, voluntary agencies, change agents and academic institutions?
- Moreover, as the political as well as the administrative structures are not truly responsive to the causes of the poor PAPs, how to revitalise their role?
- How can the PAPs have easy access to the judiciary or legal framework?

- Since the technical persons involved in designing the project usually focus on efficiency and neglect the equity and social justice aspects, the technical feasibility analysis of the project often neglects R&R aspects. So the issue is how to sensitise the technical persons to the problems of R&R?
- Even the conventional financial and economic viability studies of any development project do not pay adequate attention to R&R. There is hardly any risk and uncertainty analysis with respect to R&R. Thus, there is a need to incorporate R&R into the financial and economic analyses.
- Since planning and execution of R&R policies lack human values, the authorities always render lip services towards the pains of PAPs. How to inculcate human values in them? and
- How to maintain sustainable development in the region along side the involuntary displacement?

Photograph 5: Showing the Flock of Goats who Sustain on the Common Property Resources in the Bellary Region of Karnataka.

IV. Government's Environmental Policy and Approach towards Sustainable Development

In view of the environmental degradation owing to promotion of various development projects, such as, mega irrigation schemes, large scale industries as well as mining and mining process industries, the activists and environmentalists have raised their voices against these large projects all over the world. The old economic maxim says "grow or die" was challenged by the environmentalists. Further, in 1972 the Club of Rome's "Limits to Growth" challenged the conventional wisdom and warned that there are limits—especially environmental limits—how big human civilisation and their appetite for resources could get beyond a certain point. To add to it, the environmental awareness in the 1970s and the consequent 1st UN Conference on Environment and Development (UNCED) in 1972 at Stockholm, which was attended by most of the Heads of the Governments including our the then Prime Minister Indira Gandhi, the world body had focused on protection and regeneration of the environment, more particularly, how to ensure availability of the natural resources to the future generations. With a view to draw a strategy towards environmental regeneration and protection, it had set-up the World Commission on Environment and Development (known as the Brutland Commission). The Commission had submitted its seminal report ***"Our Common Future"*** in 1987 in which it has endorsed the concept of sustainable development (SD). It had defined SD as "meeting needs of the present generation without compromising the needs of future generations." It contains within it two concepts, (i) The concept of 'needs' of the world's poor, to which overriding priority should be given, (ii) The idea of limitations imposed by the state of technology and social organization on the environment's ability to meet present and future needs. The concept of sustainable development was adopted by the World Body in the Rio de Janeiro Conference (UNCED), also called The Earth Summit, 1992. It envisages that development becomes sustainable if and only if the stock of overall capital assets remains constant or rises over time. The natural resource base of a country and the quality of its air, water, and land represent a common heritage for all generations. To destroy that endowment indiscriminately in the pursuit of short-term economic goals penalizes both present and especially, future generations. Hence, we have to bring a balance between our development programmes and policies as well as environment protection/regeneration policies to improve the well-being of mankind. Thus, SD encompasses in an integrated way considerations of the economic growth, social progress, equity, protection of environment, and rational use of natural resources. It further calls for proper environmental accounting.

Being influenced by these international developments, the Government of India had has devised its environmental policies since 1970s and in order to protect and improve the environment, it had introduced some amendment in our constitution. The 42nd Amendment of our constitution in 1976 had injected a new dimension to public responsibility by obligating the central government to protect and improve environment for the good of the society as a whole. The policy envisaged the word "environment" in a broad sense as the quality of physical-natural environment . . . the air, water, land including forest and mines and wildlife of the country, which ought to be preserved for future generations. It provides both positive and dynamic connotations, so that state can take steps and impose restrictions on use of natural resources which adversely affect the environment. Further as per article 253 of our constitution, the parliament is empowered to make laws to give effect to international agreements and conventions (valid for all or any part of the territory of India). Notwithstanding these constitutional provisions, in our federal structure as states have the authority for ownership, management and use of natural resources such as water and land, they can make their laws and adopt policies to preserve these resources. However, under article 254, if there is any overlapping between union and state laws, the union law already has an overriding effect. And in order to enforce and enact the laws many structural changes have been introduced both at the centre and state levels. Many new institutions like Ministry of Environment and Forest, Central Pollution Control Board, State Pollution Control Boards, Waste Land Development Board have been created to take care of the environment. In addition, many Ministries and departments of the Government including Ministry of Agriculture, Ministry of Rural Development, Ministry of Metals and Minerals, Ministry of Industries, Ministry of Energy, Department of S&T, Department of Non-conventional energy, etc. are working towards protecting and regenerating the environment. No doubt, in shaping the environmental issues and objectives, our environmental thinking took its cue from the industrially advanced countries of the West.

Specially with the creation of institutional arrangements, supported by legislative measures, a number of programmes for environmental management have been implemented in our country during the last few years with focus on sustainable development. And accordingly we have perceived our programmes and policies with these broad objectives such as:

- Control of industrial pollution; and
- Preservation of the threatened species—of both flora and fauna

Subsequently, we added two more objectives, viz.

(a) Prevention of any further degradation and depletion of the country's basic natural resources and life-support systems of land, water and vegetation; and

(b) Provision for all human settlement with at least clean drinking water and a minimum level of sanitation.

It is a matter of gratification that there is a growing realization by the proponents of different projects including mining in our country regarding the importance of these four-point core environmental issues for our long-run economic development and this approach has been the corner stone of various policies and programmes. They deserve even greater emphasis for preservation of our country's production base, to combat industrial pollution and insanitation, and to preserve our socio-economic and cultural environment in the interest of public welfare. Moreover, it will help attaining the new millennium goal of sustainable development with the help of participation of all stake holders. Some of the important policies of the recent years are the Forest Conservation Act, 1980; the Air (Prevention and Control of Pollution) Act, Rules, and Amendments, 1981, 1982, 1983, 1987; the Model Regional and Town Planning and Development Act, 1985; The Coal Mines (Conservation and Development) Amendment Act, 1985; the Environment (Protection) Act and Rules, 1986; Hazardous Waste (Management and Handling) Rules,

1989; the Costal Regulation Zone Notification, 1991; the National Environment Tribunal Bill, 1992; the Environmental Audit Notification, 1992; Environmental Impact Guidelines/ Notification, 1994; Municipal Solid Waste (Management and Handling) Rules, 1999, Manual of Municipal Solid Waste (Management and Handling) Rules, 2000; New Environmental Policy, 2004; and Environmental Impact Assessment (EIA) Rule, 2006 under which apex committees are formed to examine the environmental impacts and abatement measures proposed by the proponents to contain environmental damages under Corporate Social Responsibility (CSR) for all development projects.

V. EIA and CSR as Tools for Mining and Mineral Processing Sector to Attain Sustainable Development in India

While undertaking the task of development, for the first time the concern for preserving the quality of life and protecting the environment was stressed in the Fourth Five-Year Plan (1969-70 to 1973-74). The Fourth Plan document recognized the need to introduce environmental protection concern into national planning and development. However, in the Sixth Five-Year Plan (1980-85), the concern was turned into concrete action by launching several programmes relating to enhancement of the quality of life by incorporating elements in various development projects. Adopting these norms, the Government focused on Environmental Impact Assessment (EIA), which has become a statutory requirement for all the projects in India and many EIA studies were commissioned by the Government to understand the various linkages. To add to it, the hue and cry raised by the NGOs and other voluntary organisations in the Narmada and Tehri dam projects, Silent Valley Project, Balliapal Test Range Project and the subsequent intervention by the World Bank on the resettlement and rehabilitation policies of a number of projects, as they are funded by the Bank and its associates, have brought a new awareness to our system of environmental management. In recent years rehabilitation and resettlement (R&R) issues have become a major concern for the large projects where it involves involuntary displacement of inhabitants from their ancestral hearth and homes. To overcome the losses to the physical as well as the socio-economic environment of the project affected area and project affected persons, the funding agencies and the government are insisting on EIA/SIA/SES studies to draw suitable R&R plans and abatement/remedial measures with a focus on CSR prior to the commencement of the project. The project proponents are asked to submit EMP which are to be monitored from time to time for preservation of the environment.

In view of these developments, it has been realized by the Government as well as by industry/project authorities that by undertaking the E1A studies the decision making domain improves in terms of:

- better quality of decision making in terms of immediate execution or revision or postponement of the decision;
- forecasting/predicting possible scenarios of a project;
- selecting appropriate sites for projects;
- optimum choice of resource utilisation patterns; and
- drawing of Remedial Action Plan (RAP) for project affected areas and project affected persons (PAPs).

In tune with these policy perspectives, the Ministry of Forestry and Environment now-a-days enforces that any new project needs environmental clearance in terms of three different environmental dimensions, viz., impact on flora and fauna, land, water and air pollution and impact on socio-economic environment. The socio-economic dimension implies that the impact of the project on the socio-economic and on cultural set-up has to be assessed beforehand with the help of EIA studies and to work out Remedial Action Plans (RAPs) both in the core as well as buffer zones. Under the new policy, the project proponent is supposed to spend around 2% of their turnover as CSR, CSR reporting draws much inspiration from its much older cousin, environmental and sustainability reporting. Environmental Impact Assessment (EIA) is a tool used to identify the environmental, social and economic impacts of a project prior to decision-making. Environmental Impact Assessment (EIA) is a tool used to identify the environmental, social and economic impacts of a project prior to decision-making. It also aims to predict environmental impacts at an early stage in project planning and design, and to find ways and means to reduce adverse impacts.

Scope of EIA in Mining and Mineral Processing Sector

Social and economic development in most developing countries currently stands as a dilemma between meeting basic human needs of the increasing population on one hand, and conserving declining natural resources on the other. Development projects, such as, dam, road, power line construction (just to mention a few) brings about changes in the ecosystem in which they are undertaken. Such changes affect both the environment and well-being of man. These impacts can be adverse if the processes are not well regulated or controlled through improved project selection and more responsive planning and design. Thus, project proponents have to introduce environmental impact assessment (EIA) as a tool to charter a new course of development action, which could ensure a balance between environmental protection and human development.

In tune with the objectives of NEP 2004, the Government has laid stress on conservation of critical environmental resources like forestry and minerals. Further, it has focused on intra-generational equity in terms of ensuring livelihood security for the poor, attaining efficiency in environmental resource use, and on environmental governance based on transparency, rationality, accountability, reduction in time and costs, and participation. Emphasis is given on enhancement of resources for environmental conservation too. Some of the basic principles as enunciated in the policy are:

(a) Human beings are at the Centre of Sustainable Development Concerns;
(b) Environmental Standard Setting and environmental offsetting;
(c) To undertake proper preventive action; and
(d) For strategic interventions, besides legislation and the evolution of legal doctrines for realization of the objectives.

In tune with the Environmental Policy, 2004, the Government of India has announced its EIA Policy, 2006 and the main features of the policy in relation to all projects and activities are as follows:

I. All projects and activities are broadly categorized in two categories—Category A (requiring land above 50 ha or located in a coastal zone or in the inter-state boundary) and Category B (requiring less than 50 ha and more than 5 ha of land)-based on the spatial extent of potential impacts on the human health and nature and man made resources.

II. All projects or activities included in Category A in the Schedule, including expansion and modernization of existing projects and activities and change in the product mix, shall require prior environmental clearance from the Central Government in the Ministry of the Environment and Forest (MoEF) on the recommendation of and Expert Appraisal Committee (EAC) to be constituted by the Central Government for the purpose of this notification.

III. All projects or activities included in Category B including expansion and modernization of existing projects and activities or change in the product mix but including those which fulfil the General Condition (GC) stipulated in the Schedule will acquire prior environmental clearance from the State/Union Territory SEIAA. The SEIAA shall base its decision on the recommendation of State or Union territory level SEAC as to be constituted by the Government.

As most of the mining activities involve more than 5 ha of land, the grant of any lease calls for undertaking proper EIA. By using E1A both environmental and economic benefits can be achieved, such as, reduced cost and time of project implementation and design, avoided treatment/clean-up costs and impacts of laws and regulations. The key elements associated with undertaking EIA of any mining project involve:

- *Scoping*: identify key issues and concerns of interested parties. The scope and boundary of EIA has been confined to the core zone and the buffer zone. While the buffer zone for the project requiring more than 25 ha is defined as 10 km radius from the core zone, it is 5 km radius for the projects requiring less than 25 ha of land;
- *Screening*: decide whether an EIA is required based on information collected;

- *Identifying and evaluating alternatives*: list alternative sites and techniques and the impacts of each;
- *Appraisal* of the project implies *ex-ante* evaluation of the possible impacts;
- *Mitigating measures dealing with uncertainty*: review proposed action to prevent or minimise the potential adverse effects of the project;
- *Issuing environmental statements*: report the findings of the EIA.

As regard screening, scooping, the EACs at the Central Government as SEACs at the State or Union Territory level shall screen, scope and appraise projects or activities in Category A and Category B respectively. It is also envisaged that EACs and SEACs shall meet at least once every month.

(a) Scoping refers to process by which the EAC in the case of Category A projects or activities, and SEAC in the case of Category B projects or activities, including application for expansion and modernization and/or change in the product mix of existing projects or activities, determine detailed and comprehensive Term of Reference (TOR) addressing all relevant environmental concerns for the preparation of EIA Report in respect of the project or activity for which the prior EC is sought. The EAC or AEAC concerned shall determine the TOR on the basis of information furnished in the respective application Form 1.

(b) The TOR shall be conveyed to the applicant by EAC or SEAC as concerned within sixty days of the receipt of the Form 1. The TOR suggested by the applicant shall be termed as the final TOR when approved by committee for undertaking the EIA studies. The approved TOR shall be displayed on the website of the MoEF and concerned SEIAA.

(c) The new rule focuses on public consultation which refers to the process by which the concerns of local affected persons and others who have plausible stake in the environmental impacts of the projects or activity are ascertained with a view to taking into account the material concerns in the project. All Category A and Category B projects or activities shall undertake public consultation. After completion of the public consultation, the applicant shall address all the material environmental concerns expressed during this process, and make appropriate changes in the draft EIS and EMP. The applicant to the concerned regulatory authority for appraisal shall submit the final EIA report, so prepared. The applicant may alternatively submit a supplementary report to draft EIA and EMP addressing all the concerns expressed during the public consultation.

(d) Appraisal means the detailed scrutiny by the EAC or SEAC of the application and other documents like the final report, which will include the possible impacts on environment and the abatement measures including R&R measures, the EMP and outcome of the public consultation including public hearing proceedings, submitted by the applicant to the regulatory authority concerned for the grant of EC. This appraisal shall be made by EAC or SEAC concerned in a transparent manner in a proceeding to which the applicant shall be invited for furnishing necessary clarifications in person or through an authorized representative. On conclusion of this proceeding, the EAC or SEAC shall make categorical recommendations to the regulatory authority concerned either for grant for prior EC on stipulated term and conditions, or rejection of the application for prior EC, together with reason for the same.

The appraisal of an application shall be completed by EAC or SEAC concerned within 60 days of the receipt of the final EIA report and other documents or the receipt of Form 1 and Form 1A, where public consultation is not necessary and the recommendation of EAC or SEAC shall be placed before the competent authority for a final decision within the next fifteen days.

Scope of CSR in Mining and Mineral Processing Sector

Corporate Social Responsibility (CSR) is closely linked with the principles of Sustainable Development, which argues that enterprises should make decisions based not only on financial factors such as profits or dividends, but also based on the immediate and long-term social and environmental consequences of their activities. In fact, CSR is no more considered as a philanthropy rather it is an obligation and responsibility; it is a continuing commitment of the project proponents to behave ethically and contribute to economic development while improving the quality of life of the workforce and their families as well as of the local community and society at large. It acknowledges the debt that the corporation owes to the community within which it operates, as a stakeholder in corporate activity. It also defines the business corporation's partnership with social action groups in providing financial and other resources to support development plans, especially among disadvantaged communities. Corporate social responsibility is still sometimes seen as "green wash" to clean the sins of pollution, or "white wash" to provide a facelift to the company's public image.

CSR is a concept that organisations, especially corporations, have an obligation to consider the interests of customers, employees, shareholders, communities, and ecological considerations in all aspects of their operations. This obligation is seen to extend beyond their statutory obligation to comply with legislation. The various dimensions associated with CSR are broadly divided in terms of the physical environmental, socio-economic environment, and aesthetic environment:

- The physical environmental CSR generally encompasses some dimensions like protection of flora and fauna, control of pollution of land, water and air, diffusion of environmentally friendly technologies, precautionary approach to environmental challenges, rectification of environmental damage, respecting the principle of polluter to pay the environmental costs, promotion of greater environmental responsibility, contribution to the preservation of biodiversity, reduce energy use, limit or alter material use, reduce water use, limit emissions and reduce waste and treatment of wastage before disposal. There lies a challenge, i.e., how to convert the waste to wealth?
- On the other hand, the socio-economic CSR focuses on

impact of the project on population in terms changes in demographic patterns, promotion of economic activities like employment and income generation opportunities for the people of the core as well as buffer zone, impact on production and productivity in the area, development of socio-economic and cultural activities, contribution to access for sanitation and safe drinking water, contribution to training, education and access to information with respect to essential health facilities of the community, any help to informal sector to grow in many ways and to promote infrastructure activities like construction of schools, roads, community buildings, etc. in collaboration with stakeholders on the basis of a need assessment survey (NAS). The needs of the people in the core and buffer zone are to be carefully assessed by experts in terms of *Felt needs, Perceived needs and Forced needs.* All these needs can be ranked on the basis of priorities of the local people.

- Similarly, under the social CSR the components included are contribution to equal access to health facilities, contribution to access to basic food, housing, sanitation and sufficient safe drinking water, contribution to education and access to information with respect to essential health problems in the community, promotion of other socio-economic rights like the right to work, social security, and maternity leave, to take part in cultural life, and taking up of infrastructural activities like construction of roads, bridges, etc.
- The aesthetic CSR covers activities like development of public places like parks and temples and taking suitable measures for keeping the surrounding area clean.

The various aspects of CSR those are important for the well-being of the local people are opportunities for employment and income generation opportunities, scope of increasing production and productivity, human resource development, promotion of health and hygiene, safety and risk management.

On the other hand, the critics of CSR attribute other business motives, which the companies would dispute. For example, some believe that CSR programmes are often undertaken in an effort to distract the public from the ethical questions posed by their core operations. Some that have been accused of this motivation include British American Tobacco which produces major CSR reports which is well known for its high profile advertising campaigns on environmental aspects of their operations.

VI. Concluding Remarks

In our country many Indian companies have adhered to the CSR policies in pro active manner. For instance, leading private sector player in steel production, viz., TISCO first established a social welfare scheme in 1916 to provide assistance in the rural areas surrounding Jamshedpur. The scheme is now supported by some US$2.5 million a year from TISCO and encompasses a community development program, a trust for family health and the Tata Steel Rural Development Society (TSRDS) in terms of CSR commitment of the company. The company takes a number of rural development and economic upliftment measures in the surrounding of its mines and mineral processing units. Similarly, the major public sector producer of power, the NTPC in its mission statement on CSR has stated—"*Be a socially responsible corporate entity with thrust on environment protection, ash utilization, community development, and energy conservation*". The NTPC's approach towards CSR has also been articulated in the corporate objectives on sustainable power development as stated below:

- To contribute to sustainable power development by discharging corporate social responsibilities.
- To lead the sector in the areas of resettlement and rehabilitation and environment protection including effective ash-utilization, peripheral development and energy conservation practices.

Thus, the corporate world of mining and mineral processing should use CSR via E1A/R&R in achieving the goal of sustainable development, These companies are, no doubt, capable of overcoming most of challenges and problems faced by them. In this regard, we see some new opportunities for the mining sector to grow under the umbrella of CSR. But its very success depends on adoption of stakeholders' approach with deep commitment from the top management/proponents/owners of mines. The number of stakeholders of mining projects are the Governments (Central, state and local self-governments), the project Proponent/ Authorities, the funding Agencies/Banks, the People/Citizen Groups, The NGOs/Activists and the International Organisations, who should be involved in proper implementation of environmental abatement measures like R&R for the mining projects.

REFERENCES

Acharya, S.K. and Rath, B. (2004), "Comparison between Government Forestry Project and Social Forestry Project: A Case Studies from Orissa", Boppana Nagarjuna (Ed.) "Economic Reforms and Perspective", Serial Publications, New Delhi, pp-593-621.

Adhikari, B. (2002), " Property Rights and Natural Resource: Socio-Economic Heterogeneity and Distributional implications of Common Property Resources and Management", *EEE Working Papers Series-N.3.*

Bahuguna, V.K. (2001), "Production, protection, and participation in forest mangement", *In proceddings of the Commonwealth Forestry Conference on Forests in a changing landscape.*,1-16, Canning Bridge, Asutralia: Promaco Conventions Pty Ltd

Chopra, K. and S.C. Gulati (1998) Environmental Degradation, Property Rights and Population Movements:Hypotheses and Evidence From Rajasthan (India), *Environment and Development Economics*, 335-57.

Government of Karnataka (2004) 'Mining and Quarrying' in *State of the Environment Report 2003*, Department of Ecology and Environment, Bangalore. (http://parisara.kar.nic.in/PDF/Mining.pdf, 1 August 2007).

Government of Karnataka (2005) *Economic Survey 2004-05*. Department of Planning and Statistics, Government of Karnataka, Bangalore.

Hardin, G. (1968), "The Tragedy of Commons'". *Science*, 162.

Jodha, N.S. (1986a), "Population Growth and the Decline of Common Property Resources in India". *Population and Development Review.* Vol. 11, No. 2, pp. 169-81.

Jodha, N.S. (1986b),"Common Property Resources and the Rural Poor in Dry Regions of India". *Economic and Political Weekly*. 21, No. 27, pp. 247-64.

Jodha, N.S. (1995), 'Studying Common Property Resources, Biography of a Research Project', *Economic and Political Weekly*, XXX (11, March): 556-59.

Kumar, S. (2002), "Does Participation in Common Pool Resource Management Help the Poor? A Social Cost-Benefit Analysis of Joint

Forest Management in Jharkhand, India", *World Development*, Vol. 30, No. 5, pp. 763-82.

Lipton, M. (1976), "Migration from Rural Areas of Poor Countries: The Impact on Rural Productivity and Income Distribution", *Paper presented at Research Workshop on Rural Labour Market Interactions*, International Bank for Reconstruction and Development, Washington, DC.

Ostrom, E. (1990), "Governing the Commons: The Evolution of Institutions for Collective Action," Cambridge University Press.

Pasha, S.A. (1992), "Common Pool Resources and the Rural Poor: A Micro Level Analysis," "*Economic and Political Weekly*. 27 (46), 2499-503.

Planning Commission (2006) National Mineral Policy: Report of the High Level *Committee* (http://mines.nic.in/, accessed 18 July 2007).

Planning Commission (2007) *Karnataka Development Report*, Academic Foundation, New Delhi.

Rath, B. and Sahu, N.C. Sahu (2004), "'*Revitalisation/Renovation of Common Property Resource (CPR) Potentials as an Alternative Means to Improve the Economy of Orissa*", in R.K. Panda (Ed.) "Reviving Orissa Economy: Opportunities and Areas of Action", APH Publishing Co., New Delhi, pp. 183-207

Rath, B. and N.C. Sahu (2005), "Scope of Water Harvesting System in a Hilly Terrain: A Case Study of Reconstruction of Dholdimiri MIP, Angul,

Rath, B. (2008), "*Need Assessment Study for CSR of Auraiya Gas Power Station*", *AuGPP, NTPC Ltd, Dibiapur,Auraiya, Uttar Pradesh.*

Rath, B. (2007), "*National Resettlement & Rehabilitation (R&R) Policy of the Government of India for Project Affected Persons (PAPs): Issues and Challenges* ", Proceedings of the 1st International Conference on "Social and Environmental Consequences of Coal Mining Projects in India", organized by ISM, Dhanbad in collaboration with University of New South Wales and Australian National University, Australia to be held at New Delhi, November 19-21, 2007, pp 345-56.

Saxena, K.K. and Rath, B. (1991), "Employment Multiplier Linkages of the Integrated Steel Plants in India". *The Indian Journal of Economics*, Vol. 38, No. 2, October-December, 1990, pp. 1-19.

Sreedhar G. and K. Bhaskar (2000), "Impacts of Joint Forest Management Forest Rejuvenation and Livelihoods of participant Households: Micro Level Experiences from Andhra Pradesh", *Indian Journal Of Agricultural Economics*, Vol. 55, No. 3, July-Sept., pp. 456-57.

Wade, R. (1987), "The Management of Common Property Resources: Collective Action as an Alternative to Privatisation or State Regulation". *Cambridge Journal of Economics*, 11, pp: 95-106.

Environment, Sustainable Development and Slums

Har Govind Prakash

Economic development is a result of various process and sub-processes of interaction of various economic, political, cultural, environmental and social elements. Environment is one of the important elements which are directly related with economic development. Economic development without environmental considerations can cause serious environmental damage in turn impairing the quality of life of present and future generations. Sustainable development attempts to strike a balance between the demands of economic development and the need for protection of the environment. Sustainable development aims at maximizing the net benefits of economic activities, subject to maintaining the stock of productive assets (physical, human and environmental) over time and providing a social safety net to meet the basic needs of the poor.

One of the dominating issues concerning an urban and industrial development is urban poverty. It has emerged as a major problem in recent years. It has several dimensions such as, social, economic, environmental and political and so forth. These slums are part of the urban centers and it is a worldwide phenomenon but more pronounced in the developing countries. Due to the rapid process of urbanization these are undesirable and unwanted ugly spots in the urban centers.

According to the United Nations about two-third of the urban population in South Asia lives in slums under the condition of poverty. Living conditions in these areas are harsh and services that are usually being provided by the state in regularized and authorized urban colonies are not necessarily accessible for slum-dwellers.

Slums are universal phenomenon of urban landscape. There has been a consistent increase in India's urban population. This has posed several socio-economic and environmental problems for cities in India and one among them is the rise of slums. According to the Census 2001 in India the annual growth rate of urban population is 3 percent, in large cities it is 4 percent and in the slum area it is between 5-6 percent. Consequently the number of people living in slums has increased from 9 million in 1961 to 28 million in 1981 and 40 million in 2001. Different agencies have defined slums differently and there is a regional variation in their nomenclature and connotation but has certain similarities in terms of physical appearance, characteristics and perception like congested housing, poor sanitation and inadequate supply of drinking water. Data on slums in India have been collected by various agencies like National Sample Survey Organization (NSSO); it is only the 2001 census which has made an innovative attempt to collect and compile the data on slums across the country. Slum population has been reported from 640 towns/cities of 28 states and union territories. At the state level Maharashtra (10.64 million) leads in slum population followed by Andhra Pradesh (5.15 million) and Uttar Pradesh (4.15 million). At the metro level Mumbai (5.82 million), the countries financial and film capital, has around half of its population living in slums. Delhi (1.9 million) the national capital has the country's second largest slum population, followed by Kolkata (1.5 million), Chennai (0.75 million) and Nagpur (0.73 million).

India's national slum policy is formulated against the backdrop of increasing urban slum population and shrinking infrastructure facilities. Section one will explain the concept and problem of urbanization. Section two will go through the emergence and problem of slums. Section three will look at the environment issues with reference to slums problem and section four will give suggestions for policy implications.

1. Concept and Problem of Urbanisation

Urbanization as a process and Urbanism as a culture are the concepts which are described with different connotations by the scholars of different disciplines. "Urbanization, in a demographic sense is an increase in the proportion of the urban population (U) to the total Population (T) over a period of time" (*Ashish Bose,* 1980). As long as U/T increases there is urbanization. Bergel put the concept in these words "urbanization is a process of transforming rural into urban areas."

From above definitions, the whole concept of urbanization is not clear enough as opined by Bose himself that we shall use the expression "Process of Urbanization" in a comprehensive sense and not in the statistical sense of an increase in U/T. We should understand the process of urbanization as a continuing process which is not merely a concomitant of industrialization but a concomitant of the whole gamut of factors underlying the process of economic growth and social change. Nels Anderson described the concept in a more broad sense. According to him the process denotes the migration of people from rural to urban areas, to non-agricultural occupations and to alter their life style and ideology according to modern popular urban culture, he stressed more on the socio-cultural aspect of urbanization which is called by social scientists as 'urbanism'. Urbanism is the soul of the urbanization process which symbolizes the remarkable changes in the attitudes, ideology, institutions, value-system and the patterns of social relationship. In short the change in total life-style of the people living in urban setting. Writh and Redfield have specified the concept of 'urbanism as totally contrast to the traditional rural way of life. They stated that urbanism as a way of life is typified by secularization, voluntary organizations, segmental social roles and poorly, defined norms. It is a world of tenuous relationships.'

It should be clear here that urbanization is not merely a

mechanical process of geographical or demographic mobility. It is a process of total transformation which includes geographical-demographic changes, change in the social structure, social functions, religious and value system, thinking of the people and their whole cultural sub-systems which means the radical and fundamental change in the life of urbanizing nations. Mizruchi stated that rapid and profound transformations of man's social condition in the world today is emergent of urban process. Cities provide the social context for political discussion affecting both world and country. Cities are arenas of interaction not only for political elites, but for intellectual and technological elites as well. The growth of cities has created a different type of social milieu which requires intense understanding, Meadows clarifies the complexity inherent in the concept. He said that the concept includes both the aspects—external and internal. External is the process by which rural areas transform into urban areas, urban values are diffused, movement occurs more and more towards cities and behaviour patterns are transformed to those which are characteristic of groups in cities. This external aspect is called urbanization. But it is the one side of the picture. The more important aspect is internal which is a cultural phenomenon that is called urbanism. It is a pattern of existence which deals with the accommodation of heterogeneous groups, a relatively high degree of specialization in labour, pursuit for non-agricultural occupations and development of advanced learning and art etc. Urbanism refers to a number of values which can be seen in the interaction and behavioural patterns of the people living in cities, specially in big cities. These changes related to social institutions and attitudes and values are indirect and are the repercussions of the process of urbanization which create some times, instability and disorganization in the social system. Wilbert observed that "If we accept the common sense meaning of revolution as rapid extensive fundamental social change, the social implication of industrialization and urbanization are indeed revolutionary."

Indian Census Definition of Urban Area

In Census of India, 2001 two types of town were identified:

(a) *Statutory towns*: All places with a municipality,

corporation, cantonment board or notified town area committee, etc. so declared by state law.

(b) *Census towns*: Places which satisfy following criteria: (i) A minimum population of 5000; (ii) Atleast 75% of male working population engaged in non-agricultural pursuits; and (iii) A density of population of at least 400 persons per sq. km.

Urban Agglomeration

Urban agglomeration is a continuous urban spread constituting a town and its adjoining urban outgrowths (OGs) or two or more physical contiguous town together and any adjoining urban outgrowths of such towns. Examples of outgrowths are railway colonies, university campus, port area, military campus, etc. that may come up near a statutory town or city. For census of India, 2001 it was decided that the core town or at least one of the constituent towns of an urban agglomeration should necessarily be statutory town and the total population of all the constituents should not be less than 20,000 (as per 1991 census). With these two basic criteria having been met, the following are the possible different situations in which urban agglomerations could be constituted:

(i) A city or town with one or more contiguous outgrowths;
(ii) Two or more adjoining towns with or without their outgrowths; and
(iii) A city or one or more adjoining towns with their outgrowths all of which form a continuous spread.

World Urbanisation

The urban population (UN, 1993) was estimated to be 2.96 billion (Table 1) in 2000 and 3.77 in 2010. It was estimated that nearly 50 million people are added to the world's urban population and about 35 million to the rural population each year. The share of world's population living in urban centers has increased from 39% in 1980 to 48% in 2000. The developed countries have higher urbanization level (76% in 2000) compared with developing countries (40% in 2000). The urbanization level has almost stabilized in developed countries. Africa and Asian countries are in the process of urbanization.

TABLE 1

Percentage of World Population Residing in Urban Areas by Region

World/Region	*1980*		*1985*		*1990*		*2000*		*2010*	
	%	*in billion*	%	*in billion*	%	*in billion*	%	*in billion*	%	*in billion*
World	39.4	1.752	41.2	1.997	43.1	2.282	47.6	2.962	52.8	3.779
More developed region	70.2	.797	71.5	.838	72.7	.880	75.8	968	79.1	1.060
Less Developed region	28.8	.954	31.5	1.159	34.3	1.401	40.3	1.993	46.8	2.717
Africa	27.3	.130	29.6	.164	32.0	.205	37.6	.322	44.2	.493
Asia	26.2	.678	28.6	.813	31.2	.974	37.1	1.369	43.8	1.845
Latin America	65.0	.233	68.4	.273	71.5	.315	76.6	.400	80.4	.482

Source: World Urbanisation Prospects—The 1992 Revision, United Nations, New Work, 1993.

Volume and Trend of Urbanisation in India

India shares most characteristic features of urbanization in the developing countries. Number of urban agglomeration/town has grown from 1827 in 1901 to 5161 in 2001; number of total population residing in urban areas has increased from 2.58 crores in 1901 to 28.53 crores in 2001 (Table 2). This process of urbanization in India is shown in Fig 1. It reflects a gradual increasing trend of urbanization. India is at acceleration stage of the process of urbanization.

TABLE 2

Population of India by Residence: 1901-2001

Census years	*Number of Urban agglomeration/ town*	*Total population*	*Urban population*	*Rural*
1901	1827	238396327	25851873	212544454
1911	1825	252093390	25941633	226151757
1921	1949	251321213	28086167	223235046
1931	2072	278977238	33455989	245521249
1941	2250	318660580	44153297	274507283
1951	2843	361088090	62443709	298644381
1961	2363	439234771	78936603	360298168
1971	2590	598159652	109113977	489045675
1981	3378	683329097	159462547	523866550
1991	3768	844324222	217177625	627146597
2001	5161	1027015247	285354954	741660293

According to 2001 census (Table 3), in India out of total population of 1027 million about 285 million live in urban areas and 742 million live in rural areas. Sex ratio, defined as number of female per 1000 male, for urban, rural and total India are 900, 945, 933 respectively.

TABLE 3

Population of India by Sex and Residence: 2001

India	*Male*	*Female*	*Total Persons*	*Sex ratio*
Urban	150135894	135219060	285354954	900
Rural	381141184	360519109	741660293	945
Total	531277078	495738169	1027015247	933

Source: INDCEN01, Census 2001, Office of the Register General.

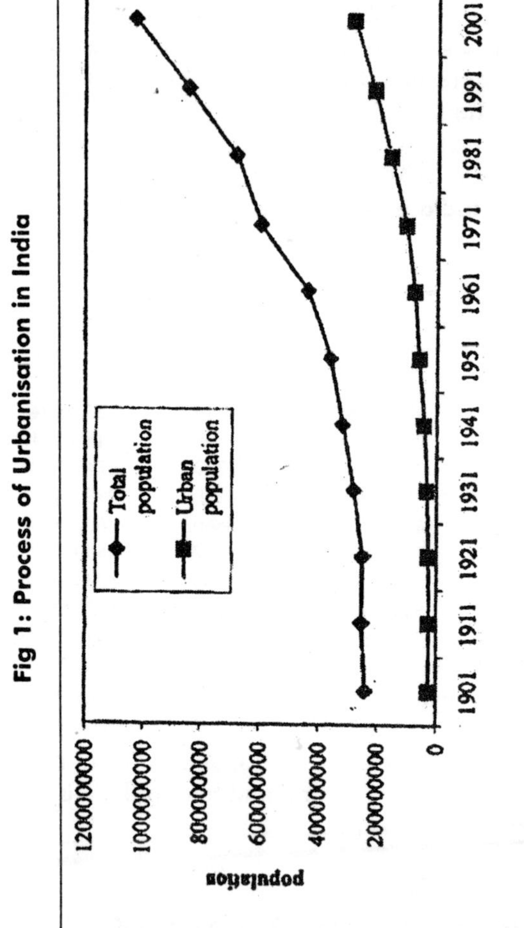

Fig 1: Process of Urbanisation in India

Degree of Urbanisation

The degree or level of urbanization is defined as relative number of people who live in urban areas. Percent urban and [U/P* 100] percent rural [(R/P)*100] and urban-rural ratio [U/R)* 100] are used to measure degree of urbanization. These are most commonly used for measuring degree of urbanization. The ratio U/P has lower limit 0 and upper limit 1, i.e. $0<U/P<1$. The index is 0 for total population equal to rural population. When whole population is urban, this index is one. The urban-rural ratio has a lower limit of zero and upper limit $\propto$, i.e. $0<U/R<\propto$. Theoretically upper limit will be infinite when there is no rural population (R=0) but this is impossible. From Table 4 it is clear that percent urban has increased from 11% in 1901 to 28% in 2001, whereas percent rural has shown gradual decrease from 89% to 72% over a century. Urban-rural ratio which is a simple index measuring number of urbanites for each rural person, experiences an increasing trend during hundred years in the process of urbanization in India. The urban-rural ratio for India in 2001 turns out to be around 38, meaning that against every 100 ruralites there are 38 urbanites in India in 2001. All these indices pin point that India is in the process of urbanization (*Sovani*, 1966) and it is at the acceleration stage of urbanization. These are presented graphically in Fig. 2.

TABLE 4

Degree/Index of Urbanisation: 1901-2001

Census Year	*Percent Urban*	*Percent Rural*	*Urban-Rural Ratio (percent)*
1901	10.84	89.15	12.16
1911	10.29	89.71	11.47
1921	11.18	88.82	12.58
1931	11.99	88.01	13.63
1941	13.86	86.14	16.08
1951	17.29	82.71	20.91
1961	17.97	82.03	21.91
1971	18.24	81.76	22.31
1981	23.33	76.66	30.44
1991	25.72	74.28	34.63
2001	27.78	72.22	38.47

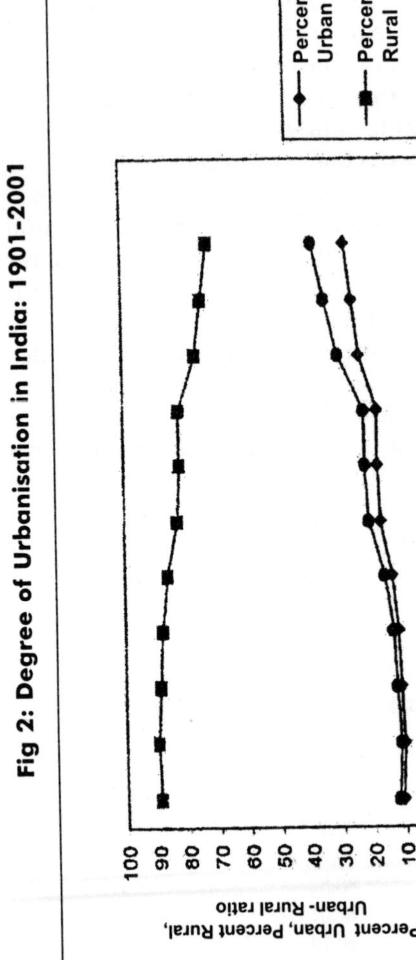

Fig 2: Degree of Urbanisation in India: 1901-2001

Urban Morphology

The following Table 5 shows number of towns and percentage of urban population by size class of city during 1901-2001. The pattern of urbanization in India is characterized by continuous concentration of population and activities in large cities (*Kundu*, 1983). This is manifested in a high percentage of urban population being concentrated in class I cities and its population has systematically gone up over the decades in the last century. As per 1901 census percentage of population in classes I, IV, V were 26%, 21%, and 20 percent respectively. According to 1991 Census, about two-third (65%) of the urban population lived in class-I cities with more than 100,000 population. In 2001 it has increased to 69%. Over the years there has been continuous concentration of population in class I towns. On the contrary the concentration of population in medium and small town (*Kundu*, 1994) either fluctuated or declined. Indeed basic reason for the increasing dominance of class I cities is graduation of lower order towns into class I categories.

From the trend in urban population by size class over the last century one can presume an increase in inequality in the urban structure, along with regional imbalance in the next decades. The distribution of population in different size class is likely to become more and more skewed. The share of class I towns or cities, with population size of 100,000 or more, has gone up significantly from 26 percent in 1901 to 69% percent in 2001. The percentage share of class IV, V and VI towns, having less than 20,000 people, on the other hand, has gone down drastically from 47 to 10 only. This is largely due to the fact that the towns in lower categories have grown in size and entered the next higher category. (*Kundu*, 1994).

Million-Plus Cities in India

Number of million plus cities (Table 6) have increased from 5 in 1951 to 23 in 1991 and to 35 in 2001. About 37% of the total urban population live in these million plus/UA cities. As per 2001 census the newly added million-plus cities are 12 in numbers, they are Agra, Meerut, Nasik, Jabalpur, Jamshedpur, Asansol, Dhanbad, Faridabad, Allahabad, Amritsar, Vijaywada, Rajkot.

TABLE 5

Number of Towns and Percentage of Urban Population by Size Class

Census Years	*No. of Towns by size class*						*Percentage of urban population by size class*					
	I	*II*	*III*	*IV*	*V*	*VI*	*I*	*II*	*III*	*IV*	*V*	*VI*
1901	24	43	130	391	744	479	26.0	11.2	15.6	20.8	20.1	6.1
1911	23	40	135	364	707	485	27.4	10.5	16.4	19.7	19.3	6.5
1921	29	45	145	370	734	571	29.7	10.3	15.9	18.2	18.6	7.0
1931	35	56	183	434	800	509	31.2	11.6	16.8	18.0	17.1	5.2
1941	49	74	242	498	920	407	38.2	11.4	16.3	15.7	15.0	3.1
1951	76	91	327	608	1124	569	44.6	9.9	15.7	13.6	12.9	3.1
1961	102	129	437	719	711	172	51.4	11.2	16.9	12.7	6.8	0.7
1971	148	173	558	827	623	147	57.2	10.9	16.0	10.9	4.4	0.4
1981	218	270	743	1059	758	253	60.3	11.6	14.3	9.5	3.5	0.5
1991	300	345	947	1167	740	197	65.2	10.9	13.1	7.7	2.6	0.3
2001	393	401	1151	1344	888	191	68.6	9.67	12.2	6.8	2.3	0.2

Class I: Greater than 1,00,000 population; Class II: 50,000-1,00,000 population;
Class III: 20,000-50,000 population; Class IV: 10,000-20,000 population;
Class V: 5,000-10,000 population; Class VI: less than 5000 population.
Source: Various Census Reports.

TABLE 6

Million-Plus Cities in India: 1951-2001

Population (in million)				*Rank*	*City*
1951	*1971*	*1991*	*2001*		
2.97	5.97	12.57	16.37	1	Bombay (Mumbai)
4.67	7.42	10.92	13.22	2	Calcutta (Kolkata)
1.44	3.65	8.38	12.79	3	Delhi
1.54	3.17	5.36	6.42	4	Madras (Chennai)
1.13	1.80	4.28	5.53	5	Hyderabad
0.79	1.66	4.09	5.69	6	Bangalore
0.88	1.75	3.30	4.52	7	Ahmedabad
0.61	1.14	2.49	3.75	8	Pune
0.71	1.28	2.11	2.69	9	Kanpur
0.48	0.93	1.66	2.12	10	Nagpore
0.50	0.81	1.64	2.27	11	Lucknow
0.24	0.49	1.52	2.81	12	Surat
0.30	0.64	1.52	2.32	13	Jaipur
0.18	0.51	1.14	1.35	14	Kochi
0.29	0.74	1.14	1.45	15	Coimbatore
0.21	0.47	1.12	1.49	16	Vododara
0.31	0.56	1.10	1.64	17	Indore
0.32	0.56	1.10	1.71	18	Patna
0.37	0.71	1.09	1.19	19	Madurai
0.10	0.38	1.06	1.45	20	Bhopal
0.11	0.36	1.05	1.33	21	Vishakapatnam
0.37	0.64	1.03	1.21	22	Vanarasi
0.15	0.40	1.01	1.40	23	Ludhiana

Basic Feature and Pattern of India's Urbanisation

Basic feature of urbanization in India can be highlighted as:

1. Lopsided urbanization induces growth of class I cities
2. Urbanisation occurs without industrialization and strong economic base
3. Urbanisation is mainly a product of demographic explosion and poverty induced rural-urban migration.
4. Rapid urbanization leads to massive growth of slum followed by misery, poverty, unemployment, exploitation, inequalities, degradation in the quality of urban life.
5. Urbanisation occurs not due to urban pull but due to rural push.

6. Poor quality of rural-urban migration leads to poor quality of urbanization (*Bhagat*, 1992).
7. Distress migration initiates urban decay.

The pattern of urbanization in India is characterized by continuous concentration of population and activities in large cities. Davis used the term "over-urbanization (*Kingsley Davis and Golden*, 1954)" where in urban misery and rural poverty exist side by side with the result that city can hardly be called "dynamic" and where inefficient, unproductive informal sector (*Kundu and Basu*, 1998) becomes increasingly apparent. Another scholar (*Breese*, 1969) depicts urbanization in India as pseudo urbanization where in people arrives in cities not due to urban pull but due to rural push.

Reza and Kundu (1978) talked of dysfunctional urbanization and urban accretion which results in a concentration of population in a few large cities without a corresponding increase in their economic base.

Urbanisation process in not mainly "migration lead" but a product of demographic explosion due to natural increase. Besides rural outjhigration (*Premi*, 1991) is directed towards class I cities. The big cities attained inordinately large population size leading to virtual collapse in the urban services and quality of life. Large cities are structurally weak and formal instead of being functional entities because of inadequate economic base.

Globalization, liberalization (*Kundu and Gupta*, 2000), privatization addressing negative process for urbanization in India. Under globalization survival and existence of the poor are affected adversely. Liberalization permits cheap import of goods which ultimately negatively affects rural economy, handicraft, household industry on which rural poor survives. The benefits of liberalization (*Despande and Despande*, 1998) generally accrue to only those who acquire new skills. It is unlikely that common man and the poor will benefit from the liberalization. Privatization cause retrenchment of workers. All these negative syndrome forces poverty induced migration (*Mukherjee*, 1993) of rural poor to urban informal sectors (*Kundu, Lalitha and Arora*, 2001). Hence migration which is one of the components of urban growth occurs not due to urban pull but due to rural push.

Problem of Urbanisation

Problem of urbanization is manifestation of lopsided urbanization, faulty urban planning, urbanization with poor economic base and without having functional categories. Hence, India's urbanization is followed by some basic problems in the field of: (1) housing, (2) slums, (3) transport, (4) water supply and sanitation, (5) water pollution and air pollution, (6) inadequate provision for social infrastructure (school, hospital, etc.). Class I cities such as Kolkata, Mumbai, Delhi, Chennai, etc. have reached saturation level of employment generating capacity (*Kundu*, 1997). Since these cities are suffering from urban poverty, unemployment, housing shortage, crisis in urban infrastructural services these large cities can not absorb these distressed rural migrants, i.e. poor landless illiterate and unskilled agricultural labourers. Hence this migration to urban class I cities causes urban crisis more acute. Most of these cities using capital-intensive technologies can not generate employment for these rural poor. So there is transfer of rural poverty to urban poverty. Poverty induced migration of illiterate and unskilled labourers occur in class I cities addressing urban involution and urban decay.

Indian urbanization is involuted not evoluted (*Mukherji*, 1995). Poverty induced migration occurs due to rural push. Mega cities grow in urban population (*Nayak*, 1962) not in urban prosperity and culture. Hence it is urbanization without urban functional characteristics. These mega cities are subject to extreme filthy slum and very cruel mega city denying shelter, drinking water, electricity, sanitation (*Kundu, Bagchi and Kundu*, 1999) to the extreme poor and rural migrants.

II. Emergence and Problem of Slum

When human beings were able to produce more than they consumed and had found ways of storing the surplus to provide for a large number of people, living away from the field, they settled on such areas which provided good environment, climate and soil favourable to plant and animal life, an adequate water supply, ready materials for providing shelter and easy access to other peoples. Concentration of population grew at the intersections of trade routes, at harbors and at the mouths of rivers with easy access to the sea. Athens, Rome were located near the sea. Mecca, Damascus and Samarkand were island cities located on caravan routes. In 1800 only 2% of world population lived in towns of more than 5000 inhabitants. No more than 45 cities had population over 1,00,000. The 19th and 20th century saw enormous growth of urban population and cities were not able to sustain the pressure of increased population and could not provide good environment and basic services to new entrants as they were unable to afford reasonable shelter within their means. They were therefore forced to live in slums.

The word slum which first appeared in Veux's Flash Dictionary in 1812 was derived from slumber which means a sleepy unknown back alley. Slum meant wet mire' where working class housing was built during British industrial revolution in order to be near the factories. These were uncontrolled settlements and lacked basic services and only poor people lived there. According to an Expert Group of the United Nations, a slum is an area that combines to various extents the following characteristics namely; (i) inadequate access to safe water; (ii) inadequate access to sanitation and other infrastructure; (iii) poor structural quality of housing; (iv) over crowding; and (v) insecure residential status.

In India almost all urban settlements face the unpleasant scenario of 'slums'. Often, this has remained a vaguely defined phenomenon. Various institutions and documents have attempted to define 'slums' largely as a measure of deficiency of basic habitat services and amenities or the absence of livable environment consisting of adequate housing and infrastructure facilities.

The concept of slums and their definition vary considerably across the states depending upon the socio-economic conditions or local perceptions prevailing in the society. There are regional differences in the names by which these slums are known in India. In Delhi, slums are commonly called 'Jhuggi-Jhonpari', whereas in Mumbai 'Jhopadpatti' or 'Chawls' are the names for slums, it is 'Ahatas' in Kanpur, 'Bustees' in Kolkata 'Cheris' in Chennai and 'Keris' in Bangalore. But physical characteristics in most of these slums are essentially the same. They are usually a cluster of hutments with dilapidated and infirm structures having common toilet facilities, suffering from lack of basic amenities, inadequate arrangement for drainage and for disposal of solid wastes and garbage. These inadequacies make the living conditions in slums extremely suboptimal, unhygienic and results in usually higher incidence of air and water-borne diseases for the dwellers.

Under Section 3 of the Slum Area Improvement and Clearance Act, 1956, slums have been defined as mainly those residential area where dwellings are in any respect unfit for human habitation by reasons of dilapidation, overcrowding, faulty arrangements and designs of such building, narrowness or faulty arrangement of streets, lack of ventilation, light or sanitation facilities or any combination of these factors which are detrimental to safety, health and morals. Thus conceptually slums are compact overcrowded residential areas (and not isolated or scattered dwellings) unfit for habitation due to lack of one or more of the basic infrastructure like drinking water, sanitation, electricity, sewerage, streets, etc.

Identification of Slum Enumeration Blocks at the 2001 Census

For the Census of India 2001, following definitions of slums have been adopted for enumeration:

(i) All specified areas in a town or city notified as 'Slum' by State, UT Administration or Local Government under any Act including a 'Slum Act';

(ii) All areas recognized as 'Slum' by State. UT Administration or Local Government. Housing and Slum Boards, which may have not been formally notified as slum under any act; and

(iii) A compact area of at least 300 population or about 60-70 households of poorly built congested tenements, in unhygienic environment usually with inadequate infrastructure lacking in proper sanitation and drinking water facilities.

Salient Features

Size and Distribution of the Slum Population

A total of 42.6 million people living in 8.2 million households have been enumerated in slums of 640 cities/towns spread across 26 states and union territories in 2001 Census. The slum population constitutes 4 per cent of the total population of the country. The slum-dwellers in the country constitute nearly a seventh of the total urban population of the states and union territories reporting slum population and 23.1 percent of the population of the 640 cities/towns reporting slums. Statement below gives population profile of the country and Chart-1 provides slum, non-slum population in urban India.

Population Profile

	Population
India	1,028.610,328
Rural	742,490,639
Urban	286,119,689
States/UTs reporting slums	283,741,818
Towns reporting slums	184,352,421
Slum	42,578,150
Non-slum (640 cities)	141,774,271

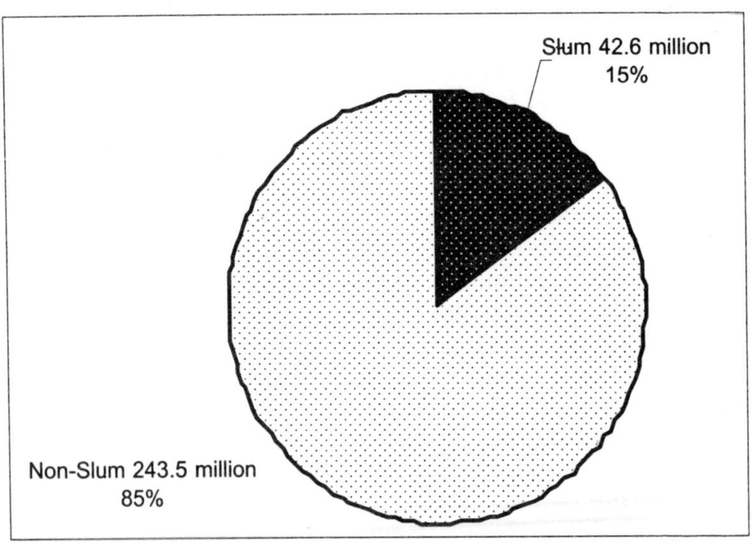

Chart 1: Slum and Non-Slum Population (in millions) of Urban India

In seven states/union territories, namely, Jammu and Kashmir, Chandigarh, Tripura, Meghalaya, Andaman and Nicobar Islands, Pondicherry and Goa, slums have been reported from less than six towns each. In the remaining states/union territories, the number ranges from 6 in Uttaranchal to 35 in Karnataka.

Slums in the 61 towns of Maharashtra account for 11.2 million people, which is more than one-fourth of the total slum population in the country. This is followed by Andhra Pradesh (5.2 million), Uttar Pradesh (4.4 million), West Bengal (4.1 million) and Tamil Nadu (2.9 million). In fact, these 5 states account for about two-thirds (65.3 per cent) of the total slum population of the country. Among other states/union territories, Punjab, Haryana, Delhi, Rajasthan, Gurajat, Karnataka and Madhya Pradesh have reported more than 1 million slum-dwellers each in the cities and towns in 2001.

As percentage of the total urban population, Maharashtra has the highest proportion of slum population (27.3%), followed by Andhra Pradesh (24.9%) and Haryana (23.2%). In thirteen States/ Union territories of Jammu and Kashmir, Punjab, Chandigarh, Delhi, Uttar Pradesh, Meghayala, West Bengal, Orissa, Chhattisgarh, Madhya Pradesh, Tamil Nadu, Pondicherry and Andaman and Nicobar Islands, 10-20 per cent of the urban population lives in slums. Kerala has the lowest percentage of slum population in the urban areas at 0.8 per cent while Goa (2.2%) and Assam (2.4%) have also a very low proportion of the slum population. Percentage distribution of the 42.6 million slum population among states and union territories is presented in Chart 2.

Kerala (2.0%) has the lowest proportion of urban population living in slums, with Goa (8.3%) and Assam (6.0%) being the only other states with less than 10 per cent of the urban population living in slums.

Slum Population in Million-Plus Cities

The large urban cities are centers of economic growth and contribute significantly to the GDP of the country. Cities with population above 100,000 account for 60 per cent of country's urban population in 2001. Rapid growth of cities in the post-independence period, however, has been associated with emergence and growth of slum and squatter settlements, characterized by overcrowding and lack of sanitation and basic infrastructure. There are 27 cities having more than one million populations in the 2001 Census in the country. Slum enumeration blocks have been delineated in all these 27 cities.

As may be discerned from the Statement 1.2 about 17.7 million population lives in slums in the cities with population above one

Chart 2: Percentage Distribution of Slum Population in States/Union Territories 2001

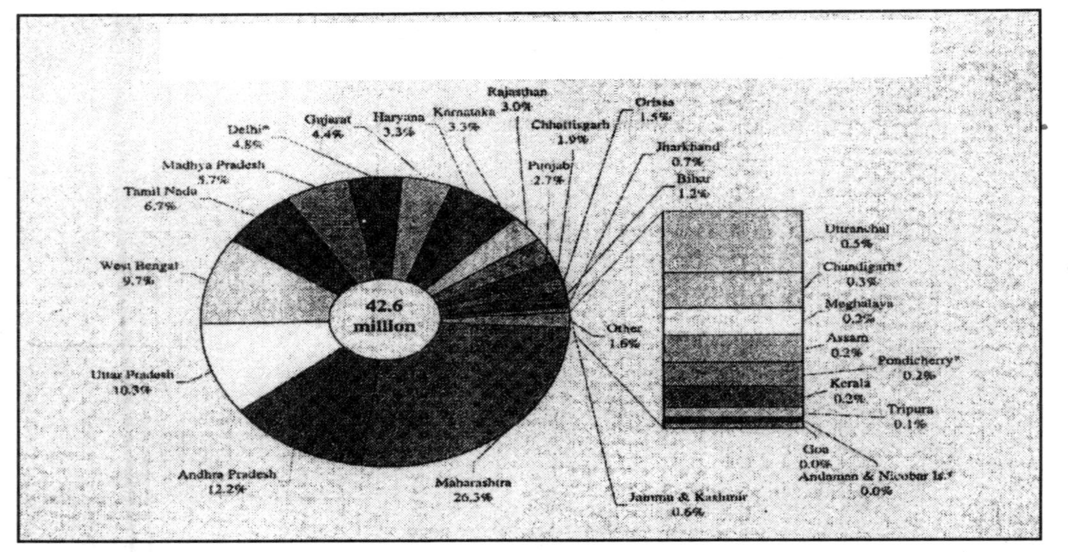

million, which is about 41.6 per cent of the total slum population in the country. In absolute numbers, Greater Mumabi has the highest slum population of around 6.5 million followed by Delhi (1.9 million) and Kolkata (1.5 million). The slum areas of Surat, Hyderabad, Chennai and Nagpur have more than half a million population each. Except Patna (3,592) and Kalyan Dombivli (34,860), all cities have slum population above 1,00,000. Chart 3 depicts distribution of slum and not slum population in million plus cities.

Four Municipal Corporations, namely Greater Mumbai, Kolkata, Delhi and Chennai together account for 25.0 per cent of total slum population of the country and around 60 per cent of the total slum population of the million plus cities as depicted in Chart 4.

Sex Composition of the Slum Population

Sex composition, i.e. the distribution of population among males and females, of slum population can be better understood in terms of sex ratio. Sex ratio of population is number of females per thousand males. There is preponderance of male population in the slum areas. The sex ratio of population in slums is 876 females per 1000 males, which is lower than that of the non-slum urban area (904) of state/union territories reporting slums. The slum areas of Meghalaya, Pondicherry and Kerala have the distinction of having more females than males. The lowest sex ratio in case of slum population has been recorded in the slums of union territory of Chandigarh (707). In the states of Jammu and Kashmir, Punjab, Uttaranchal, Haryana, Rajasthan, Uttar Pradesh, Bihar, Assam, West Bengal, Gujarat, Andaman and Nicobar Island and Maharashtra the sex ratio in slum areas is less than 900.

The Sex ratio of slum population is lower than non-slum urban population in 9 states/union territories, namely, Punjab, Chandigarh, Haryana, Delhi, Uttar Pradesh, West Bengal, Gujarat, Maharashtra and Kerala while in the remaining 17 states/union territories the sex ratio of slum population is higher than the non-slum urban areas of the respective states. Contrary to the expectation that migration to slum areas would be of males initially and is followed by their families which would give a lower sex ratio in slum areas, the higher sex ratio in slums for

Chart 3: Slum and Non-Slum Population in Million-plus Cities.

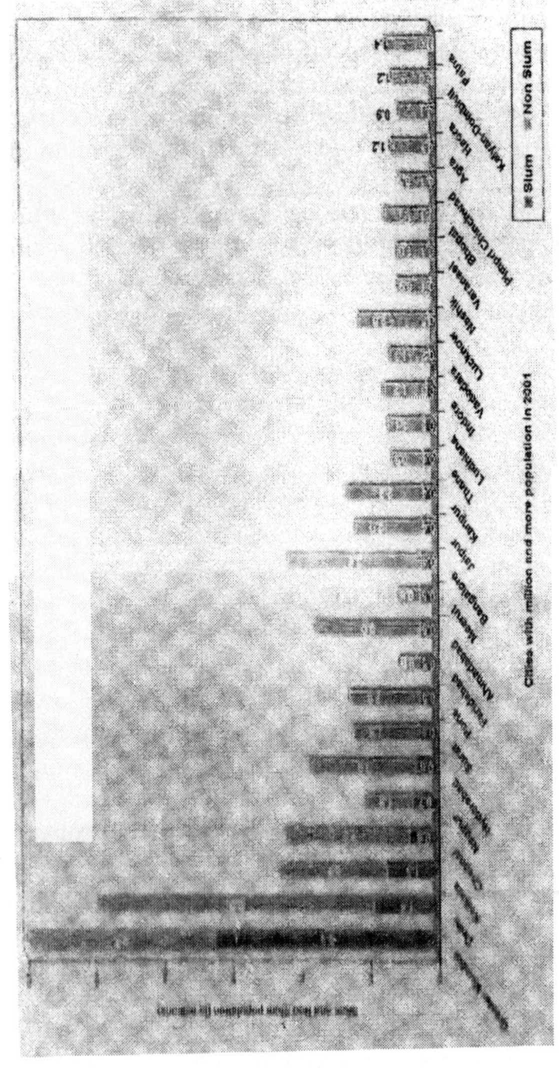

Chart 4: Slum Population in Million Plus Cities

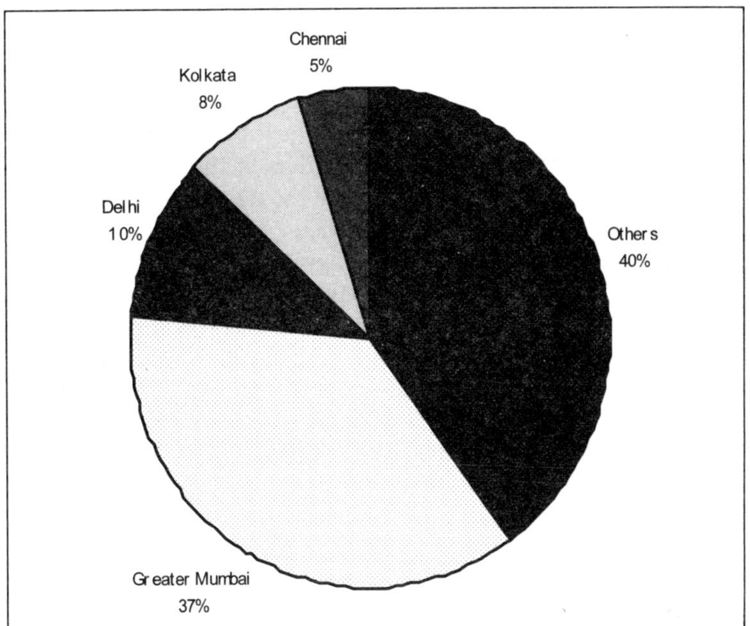

many states is suggestive of the migration to slums more often of the complete families.

In the population of 27 million plus cities, the sex ratio of slum population stands at 820 females per thousand males against 874 recorded for total non-slum population. A glance at the distribution of population by sex in the slums of million-plus cities (Municipal Corporations) reveals that Surat Municipal Corporation in Gujarat has recorded the lowest sex ratio of 701 followed by Ludhiana (759) in Punjab, Greater Mumbai (770) in Maharashtra, Haora (779) in West Bengal, Faridabad (795) in Haryana and Delhi (780). Preponderance of males in the slums of million-plus cities is evident by the fact that sex ratio in slums is less than that in the non-slum areas of the 13 cities including the large cities of Delhi, Greater Mumbai and Kolkata. Interestingly the sex ratio is more balanced in the slums of Chennai. Bangalore, Hyderabad, Pune, Jaipur, Lucknow, Nagpur, Bhopal, Patna, Agra, Varanasi, Pimpri Chinchwad and Nashik as compared to non-slum population of these cities. Kanpur has registered identical sex ratio of 857 both for non-slum population as well as slum population as well as slum population.

Literacy: Literacy is an important social indicator which throws light on the quality of life. It has long been recognized that social environment cannot improve much without attaining higher literacy levels. Hence, the prime focus here is to analyse the basic literacy pattern among the slum-dwellers in the 26 states/union territories of the country and observe the degree of inequality in the levels of literacy of the slum-dwellers.

It is noteworthy that in absolute terms only 26.7 million slum-dwellers are literate. Expectedly males are quite ahead of females in terms of literacy with 15.8 million male and 10.9 million female literates being recorded among the slum-dwellers in Census 2001. The corresponding literacy rates are 73.1 per cent for all slum-dwellers, 80.7 per cent for males and 64.4 per cent for females. The gender inequality in the level of literacy is well noticed with comparatively higher literacy rate for males and a gap of 16.3 percentage points.

Regional heterogeneity in literacy levels exists also among slums-dwellers. Overall the literacy rate in slum areas of the 26 states/union territories, which have reported slums, varies from a low of 54.8 per cent in Chandigarh to 88.3 per cent in Meghalaya. In all the 26 out of 27 States/union territories, which have reported slums, literacy rate among the slum-dwellers is lower as compared to the urban literacy rates of their respective states/union territores, Meghalaya being the only exception. In addition, Tripura, Kerala and Maharashtra also recorded literacy rates above 80 per cent. All these 26 states/union territories have registered higher literacy rate among males as compared to females. In case of males, Meghalaya is again at the top spot with literacy rate of 91.1 per cent.

Significantly besides Meghalaya, 13 states/union territories have recorded more than 80% male literacy rate among slum-dwellers. In the slums of Chandigarh male literacy rate is only 64.9 per cent which is the lowest among these 26 states/union territories. As far as females are concerned, only 3 states, viz., Meghalaya (85.5%), Tripura (81.0%) and Kerala (80.2%) have reported literacy rate above 80 per cent among slum-dwellers. Chandigarh is at the other extreme with only 39.5 per cent of its slum females as literates. The low literacy among slum-dwellers in Chandigarh is attributable to migration from area with low literacy areas.

It would be worth analyzing the male-female gap in literacy that prevails in the slums of these states/union territories. Male-female gap is highest in Rajasthan state (25.9). Meghalaya has the credit of recording the lowest gap (5.7) closely followed by Kerala (7.6). Chart 5 depicts differentials in the male-female literacy rates among the slum population in the states and union territories in 2001.

So far as literacy rates of slum and non-slum population in million-plus cities, Nagpur is the only city, which has recorded literacy rate among the slum-dwellers above 85 percent. Most of the cities fall in the range of 60 to 80 percent of the literacy rates of which four cities have registered literacy rate between 75-80 percent. Jaipur, Agra and Meerut have comparatively low literacy rate among the slum-dwellers. Patna which is at the bottom end of the list with only 52.5 per cent of its slum-dwellers being literate, also registers the lowest slum male literacy rate of 56.9 per cent. Jaipur city in Rajasthan has female literacy rate of 47.0 percent which is the lowest among the slums of the million-plus cities. As far as the differential between male and female literacy among the slum-dwellers in these million plus cities is concerned, Jaipur has registered the highest differential (26.0) while the lowest (6.3) is registered by Ludhiana.

For literacy rate of non-slum population in the million-plus cities, it may be seen that while the male literacy in the non-slum areas is higher by 7 percentage points, the female literacy is higher by almost 13 percentage points. At the city level also, the levels of male and female literacy in slum areas is distinctly lower than the non-slum population, particularly, in Patna, Agra, Meerut and Ahmadabad. Further out of 640 cities/towns, the highest number of cities/towns (249) with 38.2 percent of slum population have literacy rates in the range 70 to 80 per cent. Among the males the highest number of towns (266) has literacy rates in the range of 80-90 per cent while only 21 towns have the distinction of female literacy in this range. On the other hand, 210 cities/towns have female literacy in comparatively lower range of 60 to 70 percent. Thus the disparity in the male and female literacy rate among the slum population is pronounced in almost all cities and towns.

Chart 5: Male and Female Literacy Rates in Sulms: 2001

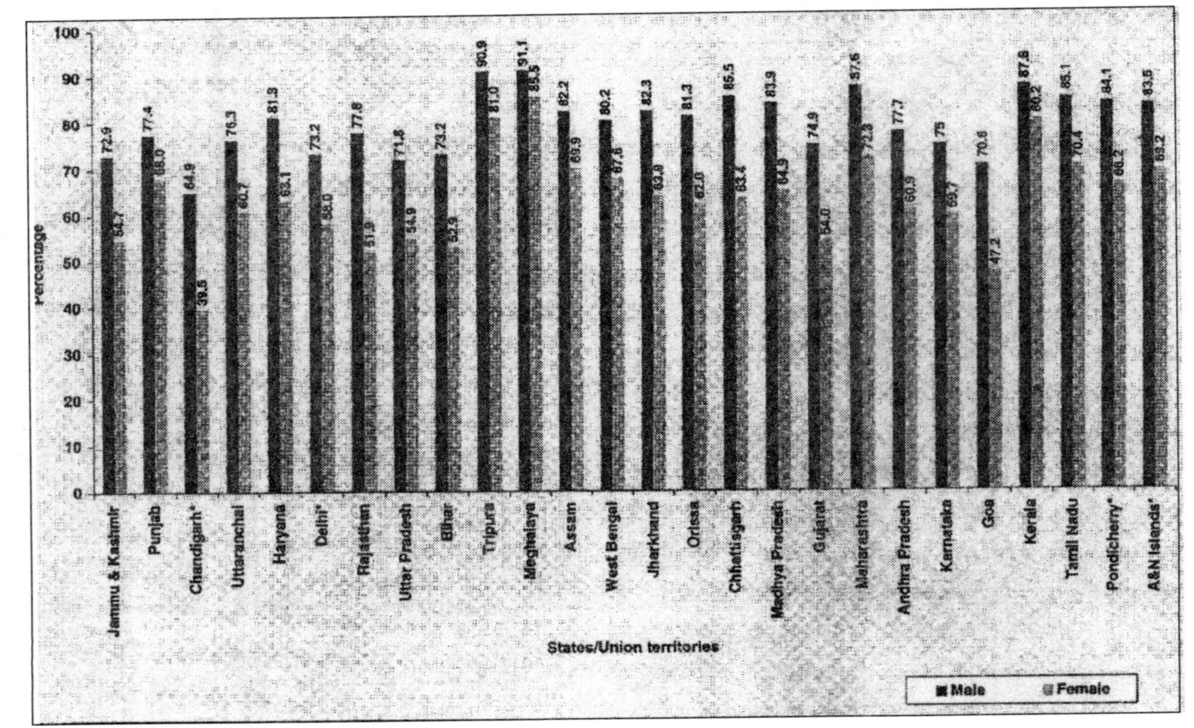

III. Environment Issues with Reference to Slum Problem

In the present century, environmental degradation has emerged as a major global concern for human survival. A large number of international and global organizations with the collective wisdom of scientists, economisists and planners have come up to settle the environmental issues confronting the national and the physical world. The environmental crisis has convinced the world to use technology and resources to repair the damage already done to our environment and also the patent substitutes for certain harmful chemicals in order to protect and preserve nature and natural resources.

We know that high degree of urbanization, high increase in population and heavy industrialization lead to many environmental and other problems such as the problem of dirty slums, air, water and land pollution, so many diseases, etc.

Rapid industrialization and urbanization have aggravated the problem of housing. It is estimated that by the turn of this century 2.5 crore additional houses will be required and by 2030, 13 crore additional houses will be required.

For the improvement of urban slums, the outlay in the Sixth Plan was Rs. 151.80 crore, while the actual expenditure incurred was Rs. 182.93 crore. In the Seventh Plan, however, the outlay was of the order of Rs. 269.55 crore, while the actual expenditure 1989 was Rs. 187.58 crore.

Nine million slum-dwellers were envisaged to be covered under the scheme of Environment Improvement of Urban Slums at an estimated cost of Rs. 269.55 crore during the Seventh Plan period, out of which 5.5 million slum-dwellers were covered up to February, 1988. Under this scheme basic amenities like drinking water supply, sewerage, storm water drains, community baths and latrines and street lightening were provide in urban slums. In spite of the huge amounts being spent on improving the conditions of the slums, the sorry state of affairs of slums is difficult to describe.

Before going deep into the problem of slums, it will be interesting to note the view expressed by commissions appointed by the government to review the problem of housing.

In the words of Royal Commission on Labour (Whitely Commission) "In the urban and industrial areas, cramped cities the high value of land and the necessity for the worker to live in the vicinity of his work all tended to intensity congestion and over-crowding. In the busiest centers the houses are built close together in order to make use of all the available space. Indeed, the space is so valuable that in place of streets and roads, narrows and winding lanes provide the only approach to the houses. Neglect of sanitation is often evidenced by heaps of rotting garbage and pools of sewage, while the absence of latrines enhances the general pollution of air and soil. Houses, many without plinths, windows and adequate ventilation, usually consist of a single small room, the only opening being a doorway, often too low to enter without stooping. In order to secure some privacy, old kerosene tins and gunny bags are used to for screens which further restrict the entrance of light and air. In dwellings such as these, human beings are born, sleep and eat, live and die.

Report of the British Trade Union congress mentions: "We visited the workers quarters wherever we stayed and had we not seen them, we could not have believed that such evil places existed. Here is a group of houses in "lines", the owner of which charges the tenant of each dwelling a monthly rent. Each hose consisting, of one dark room used for all purposes—living, cooking and sleeping, is 9 ft. by 9ft. with mud walls and loose tilted roof, and has a small open compound in front, a corner of which is used as latrine. There is no ventilation in the living room except by a broken roof or that obtained through the entrance door when open. Outside the dwelling is a long narrow channel which receives the waste matter of all description and where flies and other insects abound. Outside all the houses on the edge of each size of the strip of land between "line" are the exposed galleys, at some places stopped up with garbage, refuge and other waste matter, giving forth horrible smells, repellent in the extreme. It is obvious that these gulley are often used as conveniences, especially by the children". In the words of Jawaharlal Nehru, "These slums represent the utmost form of human degradation. Those responsible for this state of affairs should be hanged. These slums should be burned down immediately and replaced by temporary structures in more sanitary conditions." These dirty slums have adverse effects on human-beings. "While good houses mean the possibility of home life, happiness and health, bad houses spell, drink diseases immorality and crime." In the words of Radha Kamal Mukerjee, "In the thousands of slums of industrial centres, manhood is unquestionably brutalized, womanhood dishonoured and childhood poisoned at its very source."

In spite of the sincere effort on the part of government, according to one estimate, there are about 50,1,85 families homeless in Mumbai. In Kolkata, 44,440 families and in Delhi, 26,772 families are homeless.

NBO has estimated that the total number of housing units increased from 71.5 million in 1961 to 105.8, i.e. by 50 percent. Qualitatively too, there was a marked improvement. The relative share of kuccha and semipucca houses in the total stock declined in 1981 from 19 to 11.1 percent; houses increased from 46 to 64.6 percent in 1981 in the urban areas.

Most of the slums are surrounded by dirty areas; dirty (waste) water is flowing nearby and the atmosphere is not conductive to human life. Amentias of fife are also not available to the slum-dwellers. The air is also polluted and, therefore, most of the slum-dwellers suffer from one or the other disease.

According to one estimate, 25 percent of the population of Kolkata and Mumbai live in slums and dirty huts around roads and railway tracks. Similarly, in Kolkata more than 75 percent, in Kanpur nearly 66 percent and in Lucknow and Allahabad nearly 50 percent families live in one room.

In Delhi, on an average, 5 to 10 persons live in one room only. In Kanpur about 20 percent families live in kuccha house. In these slums or dirty hunts the facilities of drinking water, latrines and bath-rooms are limited. In Chennai also the same situation prevails.

In Western countries also more or less the same conditions are found in slums. In Paris, most of the slums are very old. Here, in 33 percent slums, there are no facilities of latrines, etc. Moreover, the density of population is 2.25 lakh persons per square kilometer and about 5 to 6 live in one room.

In Manchester and Birmingham respectively 20 percent and 30 percent slums are very old. Even in room dirty slums are found.

Kolkata and Mumbai are among the thirty giant cities of the world. Kolkata is often said to be the last word in urban decay and Mumbai is moving on the same disastrous path. Sandwiched between Mahim on the west and Cion on the east lies the sprawling Dharavi, Asia's, probably world's biggest slum, where nearly 3.5 lakh people live in an area of one square mile so that the total geographical area available per person is only about 60 square feet. And the per capita living space may hardly be five feet by three-less than that of a grave. Thus, dirty slums and huts are responsible for spoiling the health of the dwellers and environment in general.

IV. Conclusion and Suggestions

Urbanization is generating social and economic inequalities which warrant social conflicts, crimes and anti-social activities. Lopsided and uncontrolled urbanization led to environmental degradation and degradation in the quality of urban life through pollution in sound, air and water. Illiterate, low-skill or no-skill migrants from rural areas are absorbed in poor low grade urban informal sector at a very low wage rate and urban informal sector becomes in-efficient and unproductive.

Redirection of migration flows is required. Since the mega cities have reached saturation level for employment generation and to avoid over-crowding into the over congested slums of mega cities, it is required to build strong economic sector in the urban economy and growth efforts and investments should be directed towards small cities which have been neglected so far so that functional base of urban economy is strengthened. Then redirection of migration to these desirable destination will be possible.

Policy should also relate to proper urban planning where city planning will consist of operational, developmental and restorative planning. Operational planning should take care of improvement of urban infrastructure, e.g. roads, traffic, transport, etc. Developmental planning should emphasize on development of newly annexed urban areas. Various urban renewal process can be used. In general, urban planning must aim at: (a) Balanced regional and urban planning, (b) Development of strong economic base for urban economy, (c) Integration of rural and urban economy—emphasis on agro-based industry. Raw material should be processed in rural economy and then transferred to urban economy, and (d) Urban planning and housing for slum people with human face.

In the United Nations Millennium Declaration, world leaders

pledged to tackle immense challenge posed by mushrooming growth of slums worldwide thereby setting the specific goal of achieving significant improvement in the lives of at least 100 million slum-dwellers by 2020. This means addressing not only the needs of slum-dwellers for shelter but also the broader problem of urban poverty especially unemployment, low incomes and lack of access to basic urban services.

The following strategy may be looked into for improving living conditions in slums:

(i) Wherever *in situ* development is feasible, such slums shall be identified and taken up for *in-situ* improvement for the provision of basic facilities to make the areas habitable,

(ii) The slums located in congested and unhygienic areas of the urban centers wherein equitable distribution of space is not feasible should be cleared.

(iii) Wherever these policies are not feasible, rehabilitation and resettlement in tenaments in nearby locations should be adhered to.

The Supreme Court in a recent judgment directed authorities in Delhi to discharge their statutory obligation in keeping the city "at least reasonably clean". The court directed the concerned authorities "to take appropriate steps for preventing any fresh encroachment or unauthorized occupation of public land for the purpose of dwelling, resulting in creation of a slum". It was further ordered that "appropriate steps be taken to improve the sanitation in the existing slums till they are removed and the land reclaimed." The Court laid down the basis for this by saying that "the density of population per square kilometer cannot be allowed to increase beyond the sustainable limit. Creation of slums resulting in increase in density has to be prevented." The judges also added: "Rewarding an encroacher on public land with free alternate site is like giving a reward to a pickpocket."

Multipurpose cooperatives should raise institutional finance for improvement or construction of dwelling units and such other activity for the benefit of members, encourage habit of thrift and to make provision for the credit to needy people within the cooperative for employment generation, run ration shops, kerosene and allied commercial activities to provide daily needs at affordable cost and at their door steps, undertake educational, hygienic and other health and community-related activities, undertake repair and maintenance of common space, arrange group insurance, general insurance and other social security measures.

In India, women in urban slums live in dire poverty and are prone to violent crimes and limited employment opportunities and contribute substantially to the growth of the family. As such only the women should be admitted as members of the cooperatives. This will give status and respect to the women in the family and improve their contribution to the community. The title of the land should be given to the women members and selling of their right should be prohibited and if at all they want to leave they should surrender their right to the cooperative.

As stated above poverty is the main cause of slum formation. The multi-purpose cooperatives should undertake activities like thrift and credit, health and social services, rental housing, library, child care, care for elderly, distribution of building materials, etc. This will generate additional income for cooperatives and employment opportunities for their members.

Cooperatives should sensitize slum-dwellers about the constant threat posed by lack of basic services. Further the cooperatives should be entrusted for providing basic amenities like water, electricity, sanitary services for their members with the help of local bodies. These cooperatives will ultimately transform them into new communities wherein "each is for all and all are for each."

The HABITAT campaign is for good urban governance. It advocates transparency, responsibility, accountability, just, effective and efficient governance of towns, cities and metropolitan areas. The cities create jobs and they are strong holds of economic development and provides better quality of life to different socio-economic groups. It is, therefore, pertinent that these cities are managed properly by professionals and with people's participation. We should accept people as part of the solution and they should feel they are part of the city. In Britain, the management of the cities has been given to specialists from traffic management to garbage removal. Growing cities in Brazil, have set an example of innovative urban management by creating a sustainable urban environment and a strong sense of citizenship in its people. Housing provision is an integral part of urban development. These ideas should be implemented in Indian cities. Good governance also means involvement of residents including poor people in the decision-making of the urban governance.

There is a paramount need to invest in infrastructure to enhance economically productive activities of urban centres like electricity, access to water, sanitation, roads, footpath, waste-management, etc. This will ultimately improve the quality of life in the slums. City managers, State and national Governments should earmark substantial funds for infrastructure development thereby improving the living conditions in slums.

There is, an emergent need to discourage migration into urban centres and reverse this trend through speedy development of rural areas and small cities in order to effectively deal with mushrooming growth of slums. Instead of encouraging people to come where infrastructure is available, the infrastructure for adequate employment opportunities, better sanitation and hygienic conditions, sufficient facilities of health and family welfare, affordable housing, access to safe drinking water, transport and communication facilities, education, etc. should be made available to the poor needy people where they live.

The slums are reality of today and cities are collective future of human beings. We should take collective responsibility for their future development.

REFERENCES

Agarwal, A. (1998): Down to Earth, Nov. 30, Centre for Science and Environment.

Ali, S. (1990): *Slums within Slums*, Delhi; Council for Social Development.

Ali, S. (1998): Environmental Scenario of Delhi Slums,, Delhi, Council for Social Development.

Bhagat, R.B. (1992): Components of Urban Growth in India with reference to Harayana: Findings from Recent Censuses, *Nagarlok*, Vol. 25, No. 3, pp. 10.

Bhatnagar, B. (1991): Non-Government Organizations and World Bank Supported Projects in India: Lessons Learned, Asia Technical Department, Departmental Papers Series, No. 2, New Delhi, World Bank.

Bhushan, B. (1999): Evaluation of Accessibility and Adequacy of Basis Amenities in Urban Slums and EWS bustees: Policy issues and Agenda for Future, Delhi Ministry of Urban Affairs and Employment.

Bose, Ashish (1980): *India's Urbanisation,* Tata McGraw-Hill Publishing Company, New Delhi.

Bose, Ashish (1970): *Urbanisation in India,* National House, Mumbai.

Breese, G. (1969): *Urbanisation in Newly Developing Countries.* Prentice Hall, New Delhi.

Brockerhoff, M. (1999): 'Urban Growth in Developing Countries: A Review of Projection and Predications', *Population and Development Review,* Vol. 25, No. 4, pp. 757-78.

Davies Kingsley and Golden, H.H. (1954): "Urbanization and development in pre-Industrial Areas", *Economic Development and Cultural Change,* Vol. 3, No. 21.

Davis Kingsley (1962): Urbanisation in India—Past and Future", Turner, R. (ed.), *Indian Urban Future,* University of California Press, Berkley.

Despande, S. and Despande, L. (1998): "Impact of Liberalization of Labour Market in India: What do facts from NSSO," *50th Economic and Political Weekly,* Vol. 33, No. 22, pp. 21-31.

Khosla, R. (1998): "Problems in Slums", *Urban Poverty* A Quarterly Newsletter of the National Institute of Urban Affairs, July December 1998.

Khosla, R. (1997): "Slums and Infrastructure", *Urban Poverty*: A Quarterly Newsletters of National Institute of Urban Affairs, Jan.-March 1997.

Kundu, A. (1983): "Theories of City Size Distribution and Indian Urban Structure—A Reappraisal," *Economic and Political Weekly,* 18(3).

Kundu, A. (1994): "Pattern of Urbanization with special Reference to Small and Medicine Towns in India," Chadha, G.K. *Sectoral Issues in the Indian Economy,* Har-Anand Publications, New Delhi.

Kundu, A. and Gupta, S. (2000): 'Declining population mobility, Liberalization and Growing Regional imbalances—The Indian Case'. Kundu, A. (ed.) *Inequality, Mobility and Urbanization,* Indian Council of Social Science Research, Manak Publication, New Delhi.

Kundu, A., Baghi, S. and Kundu, D. (1999): "Regional Distribution of Infrastructure and Basic Amenities in Urban India—Issues Concerning Empowerment of local Bodies", *Economic and Political Weekly,* 34(28), July 10.

Kundu, Lalitha and Arora (2001): Growth Dynamics of Informal Manufacturing Sector in Urban India: An Analysis of Interdependence, Kundu, A. and Sharma, A.N. (eds.), *Informal Sector in India,* Institute of Human Development, New Delhi.

Moser, C. (1997): Household Reponses to Poverty and Vulnerability, Urban Management Programme Policy, Paper 21, Washington D.C; World Bank.

Mukherjee, Shekhar (1993): *Poverty-Induced Migration and Urban Involution in India: Cause and Consequences,* International Institute for Population Sciences, pp. 81-91.

Mukherjee, Shekhar (2001), Linkage between Migration, Urbanization and Regional Disparities in India: Required Planning Strategies. IIPS Research Monograph, Bombay, pp. 1-226.

Narayan, D. (1995): Designing Community Based Development, Environmentally Sustainable Development, Participation Series paper No. 007, Washington D.C. World Bank.

Nayak, P.R. (1962): "The Challenge of Urban Growth to Indian Local Government" for Turner (ed.) *op. cit*.

Parker, R., Kreimer, A. and Munasnighe, M. (1995); *Informal Settlements, Environmental Degradation and Disaster Vulnerability*, Turkey, World Bank.

Pathak, P. and Mehta, D. (1995): Recent Trend in Urbanization and Rural-Urban Migration in India: Some Explanations and Projections, *Urban India*, Vol. 15, No. 1, pp. 1-17.

Premi, M.K. (1991): "Indian Urban scene and its further Implications," Demography India 20(1).

Rawal, Y.P. (1998): Slums in the Capital City, Delhi, Government of Delhi.

Registrar General (1991): Censuses of India: Emerging Trends of urbanization in India, Occasional Paper No. 1 of 1993, Registrar General, New Delhi.

Registrar General, 2001: Census of India, 2001, India, New Delhi.

Schubeler, P. (1996): Participation and Partnership in Urban Infrastructure Management, Urban Management Programme Policy, Paper 19, Washington D.C., World Bank.

Sen, A. and Ghosh, J. (1993): Trends in Rural Employment and Poverty Employment linkage, I LO. ARTEP Working Paper, New Delhi.

Sovani, N.V. (1966): *Urbanization and Urban India*, Asia Publishing House, Bombay.

World Bank (1992), Development and the Environment, World Development Report, 1992, Washington, D.C., World Bank.

World Bank (1996): *World Development Indicators*, Washington D.C., World Bank Publication.

World Bank (1997a), *India—1997 Economic Update: Sustaining Rapid Growth*, Delhi.

World Bank (1997b), *India—Achievements and Challenges in Reducing Poverty*, Washington, D.C.

SEZ, Pollution and Sustainable Development

Anath Bandhu Mukherjee

SEZ has now turned out to be a new *Mantra* to exploit the poor nations and cause irreparable damage to their eco-system and environment. The concept of SEZ (special economic zone—nick named—special exploitation zone) is the new instrument in the hands of the developed countries to fool the politicians and market crazy economists as well, of the third world. It has been advocated as the only means of quick or rapid economic growth. Those who believe in the principle 'growth at any cost'—heavily relied on this uneconomic, undemocratic and environment damaging dope. But its experiment and trial in some countries of the world has made it clear that it is neither economically viable, nor politically desirable and environmentally fraught with the great danger of devastation

The supporters of this SEZ policy put forward three reasons for creating SEZ in a country. First SEZs usher in and ensure quick economic growth and high growth rate. Secondly, SEZs create and expand job opportunities and job market. Thirdly, they develop and increase export of goods and durables of a country to a great extent. Fourthly, the SEZs ensure and increase in the purchasing power among a large section of the people in the country. We will take up these arguments in succession. Yes it is a fact that SEZs ensure rapid economic growth and growth rate of a country. How? For making special economic zones successful—one has to allow a number of concessions to the investors / producers in the SEZs. The area has to be made a land of no control or an alien land. This area has to be kept out of the bounds of the country's laws—civil, criminal, trade union and environmental. The investors would have their own rules and laws to control everything in these areas. Not only that the products in these zones would not be subject to any kind of tax ! Infrastructure of those zones are to be developed by the state—but it would have no control on these areas and the investors and hence the SEZs would create no huge job opportunities but only a handful of jobs because the investors would use highly labour saving technology—thus giving a lie to the claim of creating "mass" employment. Only highly skilled workers would get employment. Secondly, the SEZs would, at the outset, create goods at a minimum cost—but soon the workers would find it difficult to work in the SEZ area. Because pollution would make the area not habitable for human beings. This has happened in China, Brazil and some other countries where the SEZs were used as a weapon for quick economic growth. Thirdly, SEZs have devastated the environment of these areas. Rivers and other water bodies have been polluted—making them unusable. So, the claim or achieving sustainable growth through SEZ policy is a damn lie. SEZs should have no place in the economic development of a country.

For sustainable development—environment friendly or eco-friendly agriculture and industry need be developed and encouraged so that the beautiful earth may help future generations to live here happily and freely.

SEZ or Special Economic Zone has now being put up to be a new *Mantra* to exploit the poor nations and cause irreparable damage to their eco-system and environment. The concept of achieving quick or rapid economic growth through SEZ (nick named—Special Exploitation Zone) is the new strategy devised by the developed countries and a few market crazy economists to fool the people of the third world countries for looting their natural resources. This strategy or rather policy is being advocated unabashedly as the only means to gain quick economic development and growth. Those countries which believed in the principle, "growth at any cost"—relied heavily on this uneconomic, undemocratic, socially and ecologically damaging 'pill' of the American neo-colonialists. The SEZ policy and its subsequent experiement, carried out, with much pomp, hype and propaganda, in some third world countries has made clear that it is neither politically desirable, economically viable nor environmentally sustainable. For sustainable development there is no short-cut way or method which can be used for the benefit and well-being of the people of this planet, wherever this strategy has been used, or experimented, whether in Latin America or Africa or Asia—the result has been disastrous economically but its impact on environment and ecology of these countries, has been devastating. The SEZ has been found to be the worst offending unit to pollute—land, water, air and atmosphere of the experimenting countries. In this paper these issues have been analysed and discussed. In Section I—the history of the SEZ has been taken up followed by its polluting capability, in Section II and in Section III sustainable development in relation to SEZ has been analysed and in Section IV conclusion has been drawn.

I. The History of the SEZ

A. Washington Consensus: What it is?

Nowadays most economists are well aware of these two infamous words Washington Consensus. Who were the parties to arrive at such a consensus and for whom? The parties were none other than the IMF, The World Bank and the U.S. Treasury. And for whose interest? Ten Latin American Countries led by economically devastated Argentina (Brazil, Chile etc.) joined hands and met in Washington to negotiate a despicable, uneconomic, purely political and an anti-national agreement in the 1990s. Why? To help the shattered economy of Argentina and other Latin American countries. Yes, this is partly true. But the real truth is that during the 1980s the USA and its allies were reeling under serious recession—so in 1990s, under the leadership of the USA a conference with 10 Latin American countries was held at the Institute of International Economics. In that meeting a ten point bail out programme was adopted unanimously. This is known as Washington Consensus and the most notable of the ten points were: (1) Drastic reduction in government spending, (2) Lessening of tax burden, (3) Privatisation of public sector enterprises, (4) to relieve government of development responsibilities, (5) Removal of restriction on foreign investment, (6) Reduction of subsidies on agriculture and food articles, (7) Keeping government off welfare activities, (8) Capital market liberalization, etc. and so on. The architect of this consensus was John Williamson.

Many of the ideas incorporated in the Washington Consensus were developed to cope with the specific problems of the Latin American countries and at the same time to find a way out of the American recession of the 1980s. The proponents of the consensus took pleasure in the thought that they prescribed 'right' policies for developing countries—that 'signalled a radically different approach to economic development and stabilization. The Keynesian orientation of the IMF, which emphasised market failure and the role of government in job creation was replaced by the free market *mantra* of the Washington Consensus. The ideas that were developed to cope with problems arguably specific to Latin American countries subsequently deemed applicable to countries around the world. Capital market liberalization has been pushed despite the fact that there is no evidence to showing that it spurs economic growth. (*Stiglitz*, 2002).

B. Effect of Policies set by Washington Consensus

The net effect of the policies set and imposed by the so-called consensus has been beneficial to the few at the expense of the many, the well-off at the expense of the poor. In many cases commercial interests and values have superseded concern for the environment, democracy, human right and social justice.

C. The Concept of SEZ (Special Economic Zone)

The special economic zone idea came from the said consensus. Argentina and Brazil and some other Latin American countries implemented this SEZ policy in their countries with disastrous results. We will show this here briefly. To spur growth or achieve growth at a quick pace the idea of SEZ was implemented. What then this special economic zone is? SEZ is an area, "piece of land earmarked for foreign investment and specially made for export-oriented industries. The law of governance of these places are not to be governed by the law of the land. The criminal, civil, trade union, industrial and above all environmental laws of the respective country will not be applicable in these zones! Rules and regulations regarding the setting up of SEZs will be discussed a little later.

Setting up SEZs by third world countries for achieving quick economic growth rate has ultimately been confronted with serious social, political, environmental hazards and ecological imbalance. Apart from these stakes the SEZs have not been able to improve the economic condition of those countries. But before going into the details of the social, political, economic and environmental impact of the so-called device of economic growth and development, we shall take a look at the laws, rules and regulations that usually operate in SEZones in a country.

D. SEZ Laws: Rules

Special Economic Zones have been rightly called the Special Exploitation Zones. Indeed these zones have been set up only to allow unbridled loot, exploitation of land, water, air and environment not only by the foreign multinational corporations but also by the local investors or multinationals of the country. Close look at the laws, rules and regulations will make the point and arguments clear. The laws help the multinational corporations to exploit the poor nations legally!

E. The Laws—Indian Laws: Regulations: Applicable to all other Nations

In India the investors or industrialists (multinational corporations) in the Special Economic Zones are exempted from paying (a) 22 kinds of cess (industrial); (b) all forms of excise duties and export duties/levies (Sec. 26); (c) no Service Tax to be charged on the product; (d) no Sales Tax of any kind; (e) no tax to be paid on sale and buying of shares, if these shares are bought by the foreigner or foreign company through the SEZ market;

(f) before gaining approval for SEZ all state governments should announce the maximum possible tax concessions to be allowed to the investors; (g) the SEZ investors/industrialists shall have the right to build captive power plant—transportation and distribution of electricity will also rest with them; (h) SEZ developers shall not have to pay any taxes on total profit for the first 10 years, i.e. profit is not initially taxable; (i) there will be no tax on profit of the industrialists on exports or on services for five years; (j) industry will not be bound by foreign trade rules; (k) no tax on that part of profit/to be given as dividend to the share holders; (l) the interest, banks, in the SEZ area, will pay to the foreigners will not be taxable; (m) if anyone wants to transfer business in the SEZ area he can do so; (n) any institution can operate insurance business in SEZ; (o) Banks and share market in SEZones will enjoy tax holidays for first five years to the extent of 100 percent, for the next five year 50 percent and for the next 5 years 25 percent and only after 20 years small amount of tax would be levied on the investors/industrialists.

From the above paragraph it can be seen that foreign investors/multinationals have been given free hand to loot and plunder the host country. It is however not the level play ground. Multinationals are allowed to take away huge amount of money from the host country—at the expense of nature and environment because no law or rules of the country will be applied in the SEZones. And the effect of the SEZs on the national economy has always been highly catastrophic. Some examples are shown later in this article.

E. The SEZ: Arguments for and Against

It is claimed that SEZs would quicken economic development particularly the growth rate. Secondly it will help job creation and opportunities. Thirdly, it will help growth through export and foreign trade. Fourthly, SEZ will open up ways for ancillary and complimentary industries leading to 'mass' employment. Lastly, the SEZs will help economic revival through foreign direct investment and capital movement. Let us examine these arguments. In the first place—it must be admitted that the SEZ will surely quicken economic growth rate as had happened in Brazil and China. But this growth did not last even a decade because the SEZones polluted the atmosphere of the entire zone so badly that production in those areas had to be stopped. Secondly, the claim of mass employment is a tall claim. The reason is very simple. Multinational corporations and the indigenous industrialists—now in the 21st century—will not use labour-intensive technology—instead they as logical—would employ highly capital intensive technology so that profit could be maximised. So, there is no possibility of 'mass' employment. Highly qualified and skilled persons would get employment in these special zones. Thirdly, the social cost of the special economic zones is prohibitive. Why? The laws allow the multinational corporations both foreign and national to grow more production at any cost and giving a good-bye to ecology and environment. Since the social impact and environment impact 'assessment are not enforced—so there is no way of stopping them in their attempt to denude nature and disturb ecological balance of the country.

II: SEZs IN BRAZIL, ARGENTINA, CHINA AND INDIA

A. Argentina

Before Chinese experiment with SEZ policy, Argentina and Brazil had the bitter experience of SEZ economics. This policy of SEZ was the outcome of the Washington Consensus. The 10 point programme was taken up to kill two birds with an undemocratic, anti-national and uneconomic but purely political i.e., one stone globalization. The seeds of this idea were there in the Washington consensus. It was to tide over the severe economic crisis of the 1980s of the Latin American countries on the one hand and on the other hand to save the capitalist economic system which was reeling under severe economic stagflation and inactivity seeking desperately, a way out, planned the so-called consensus.

The impact of the new economic package on Argentina was discriminating and damaging. The new consensus made their economy almost bankrupt. Inflation touched 3 digit figure unemployment rate went up to double digit (figure). There were serious problem on liquidity—interest rates were very high and bank loans were not easily available. Argentina and the other nine states thought that the consensus arrived at with so much skillful handling by the IMF-World Bank and the US Treasury would lead them to a new dawn of economic recovery.

Argentina bore the burnt of an unplanned investment, uncontrolled 'market economy' and a deceptive policy of neocolonial economic assault. With a view to implementing the IMF directed structural reforms (principles) Argentina went on a privatisation spree in an alarming speed. The government took the lead in privatisation race—land, roads, highways, gas, oil and natural resources, mineral resources were sold at a premium. Why—because disinvestment and privatisation were the two preconditions for foreign direct investment. And through these transfer of assets—flew in swarms of foreign direct investment—and as a result the economic growth rate jumped to 4 percent—which was highest in the post-war history of Argentina. Buenos Aires—capital of Argentina literally took a radical change overnight with American food chain, shopping plaza, malls and multiplexes. Special economic zones were the only word to them which stood for rapid economic growth. None cared for the huge loss of life, loss of employment, mass eviction from cultivable land, home and hearth. There were seldom any protest against all these forms of social, political and economic injustice. A group of middle class intellectuals and the media sang the paean of 'market economy' in high pitch and they predicted that this "prosperity will surely trickle down to all the sectors of the economy". This wave of Free market *Mantra* increased the inequality gap between the rich and poor, and the percentage of unemployment never came down below 12 percent—yet the drum beaters of 'market' economy highly applauded this 'growth' as a distinct sign of 'prosperity. But alas! This dream of prosperity had burst like soap bubbles within five years. The FDIs flew away followed by the South-East Asian crisis and these two together pricked the ballon of Argentina's economic prosperity and dealt blow to the proponents of 'market' economy. The situation in Argentina worsened so much that the President had to flee the country—because of the people's movement for security, safety, job guarantee and a decent living condition. Too much dependence on export and foreign investment can neither ensure economic prosperity nor economic stability.

B. Brazil: The Chemical Hub

Brazil witnessed the devastation of environment and ecological balance in setting up chemical hub in the Special Economic Zone. It was in the latter part of 1970s chemical hub was set-up in Cubatao. It became the "chemical capital of Brazil." To ensure rapid economic growth hundreds of investors were invited to open industrial units in Cubatao. This industrial capital was built up on unusual flow of foreign capital. The city was situated very close to a river—and not very far-off the most famous port of Latin America, Santos. The infrastructure was ideal—both the roadway and the waterways were excellent there. Over and above that the government allowed free or liberal industrial policy along with cheap labour, huge tax relief and land at a throw away price. So Brazil down with 3 digit inflation announced maximum quantum of concession to the neo-colonial hawks (multinationals). No other country in the world could think of allowing so much relief and concessions to the foreign industrialists. What happened then? The lure of earning maximum amount of profit turned out Brazil to be the ideal destination for international investment

(1) The Valley of Death: Cubatao

Within a brief span of a decade (in 1985) the area became a "valley of death". Uncontrolled trading, unplanned investment and allowance of abundant tax cut and relief-based export cannot help a country achieve sustainable growth. The poisonous smoke, chemical waste, non-degradable toxic element, and acids had wiped out dense Atlantic forest range of the adjacent mountain. A living rippleless river had turned into a dead cess pool. There is no life in the bed of the river—only poisoned water and a few water plants were to be found there. Land, water and air—atmosphere of the city have been polluted so much that mothers are giving birth to headless, handicapped children. The air has become heavy—and there is heavy shortage of oxygen—no one can breath healthy air. The morning sky remains clouded with dark clouds. Inhabitants suffer from, tuberculosis, cancer, asthma and various lung ailments. Frequent acid rains make life unbearable here in this city. The valley has gone out of control two decades ago—and even after a lapse of a quarter of a century and inspite of the efforts of many a country—environmental and atmosphere pollution level could neither be brought to tolerable limit nor the "valley of death" could be made habitable for human beings. SEZ has thus turned a nice valley with all the bounties of nature into a veritable hell not fit for human habitation. And so the question of ensuring economic growth and sustainable development with SEZ format is a far cry.

C. SEZ in China and its Impacts

Now we will take up the Chinese SEZ issue. China is said to be following a policy which is an amalgam of socialism and market economy. But in reality it has got nothing to do with socialism—virtually it is an attempt to bring back capitalism. They call it market socialism. The Chinese economy has by now (2009) become the third world economy in GDP measure. Only two nations are ahead of them—the USA and Japan. But China has been following the capitalist path of development since 1979—when Deng Tsiang Ping took control over the CPC. China's motto of economic development was 'Development at any cost'. The Chinese economy relied more on SEZ policy for quick economic growth. To achieve this they created many SEZs-covering 36,800 sq.km of coastal fertile land. Uncontrolled and unplanned investment by foreign multinational corporations—enabled them to earn unlimited profit at the expense of the poor Chinese farmers and marginalised people. After showing economic growth rate of 25 per cent (?) for over a decade, now the sky of Shenzen is covered with poisonous suffocating smoke. According to the United Nations Environment Division—Shenzen is such an area where environment has been damaged very quickly. Workers are being fleeced mercilessly and the crime chart is ascending at a galloping stride. Workers' wage here is US$ 80 per month and these workers have to work for over 12 to 16 hours a day! The multinational corporations mostly employ women workers in these SEZs and they are even denied the right to attend to the call of nature more than once. On the other hand, they are advised to use "toilet bags"!

Cost in Terms of Money

What has actually happened to Chinese economy for achieving her eye catching growth rate? To produce US one dollar net domestic product (GNP) China has been using 4.3 times more coal and electricity than the USA and 11.5 times more than that of Japan, one-fifth of China's population has been compelled to live in the most polluted area of the country. The multinational corporations, 60 percent of them never cared for environment or ecology of the land. A World Bank report states that three lakh Chinese are dying every year owing to environment pollution. And the Chinese government has recently admitted an annual loss due to pollution of US$ 200 billion which is 20 percent of its Gross Domestic Product (GDP).

Social Cost of Rapid Growth

In China anti-eviction riots and crimes of various types are increasing every year at an alarming rate. In 2004 China had to face 74,000 anti-eviction riots. In 2007 the number of riots and other criminal activities rose to 80,000. Shenzen records crimes almost 9 times more than Shanghai's crime rate. Retrenched women workers are forced to go for prostitution. China is the home for 18 per cent of world's poor population. In China per capita land is 0.094 hectare. And during the 13 year (1992-2008) a little over 20 million farmers have been evicted from their land.

Measures Chinese Government Prescribed: No More Development at any Cost

The Chinese Government took a decision in April 2004 that permission would not be given to use agricultural land for non-agricultural purposes. The Chinese agriculture minister had declared that core area of agricultural land would not be reduced. China has now drastically reduced the area and number of SEZs ones; Zinhua reported, in 2003 there were 6866 SEZs but at the end of the last year the number of SEZs have come down to 1568 and the area under SEZ has now been reduced to 1949 sq.km. i.e., 77.16 percent SEZs have vanished now. The State Council has decided not to allow permission to create SEZs near wet land and forest area for the protection of environment. "The Chinese Communist party in its Congress (2007) were serious to frame up policies to wean the economy off its dependence on exports and to expedite its climb up technology ladder." The President Hu Jin Tao said, "China was determined to shift from growth at any cost trajectory to a more balanced and sustainable model of development. China's emphasis on environmental problems should be taken seriously. She has so far destroyed her environment in several regions. The rat race of development at any cost has landed China in a cesspool of quagmire."

We will just recollect a very important instrospection of a Chinese minister. Pan Yul, a vice-minister of China's State Environmental Protection Administration spoke in 2005, "China's rapid development often touted as an economic miracle has become an environmental disaster. The Chinese leaders will soon know that economic miracle will end soon because the environment can no longer keep pace with development goals. China will have by 2010, nearly 130 million cars and by 2004 more cars than the USA." China draws 70 percent of its energy needs from coal and that it typically uses six to seven times more energy to produce a dollar of output than do developed countries, the extent of calamity that may engulf China and the world becomes clear. And one has not even mentioned the problems posed by rampant desertification, polluted rivers, depleted ground water reserve and what not." This SEZ economy has, widened the inequality gap in the Chinese society, widened the lopsided development of the coastal area and the mainland. It has created social rift in the Chinese families. It has accentuated the unemployment problem. This policy has landed China into a vortex of vicious pollution circle.

So the question of sustainable development is a myth to these protagonists of rapid economic growth. China, though—very late—still has now woke up to the need of the hour. But what about India? We will see now.

D. Indian SEZ Policies and Practice

WTO led or influenced globalisation policies suggest or advise the third world countries should benefit much by switching over to non-food commercial crops that cater to the requirement of fast growing middle and upper middle class households. This is going to endanger food security of the poor people. India during 1990-91 and 2001-02 has lost 60 lakh hectare agricultural land. Now—what is produced in those areas? Non-food products, e.g. fruit and flower and fodder for animals are being produced there. As a result, the rate of food grains production is lagging behind the population growth. This means we shall have to buy food grains from outside: India attained a somewhat self-sufficiency in food grains production in the eighties of the last century and this achievement has benefited our country very much. We no longer beg food grains from the West to feed our hungry millions. But this situation is changing very rapidly—because of the manipulative policies of the WTO. India's position as a food surplus country has been eroded of late. Last year India had to import a few lakh mt of wheat and rice from the developed countries. This is an ill omen for Indian economy.

India is now following China's SEZ strategy as a way of economic growth. India started with EPZ in the different parts of the country. But now India has allowed so far 550 SEZ—spread over different states. The area under special economic zones is nearly 125,000 hectare. The special economic zones which are operational here—have not been able to create 'mass' employment. The reason is—the technology which is being used in the SE Zones are highly capital intensive. So, the SEZs cannot create 'mass' employment—it will employ only a few highly technical persons and not the unskilled unemployed 'masses'. Secondly, these SEZs will not produce goods for domestic consumption. SEZs will produce goods and articles for export and for the developed countries.

Thirdly, some of the SEZs are being used to produce articles—which are banned in the developed countries. Specially the chemical hubs are set-up not to help or quicken economic growth of the country—but to cater to the needs of the greedy people of the developed countries. From these arguments it is clear that economy of SEZ, as a means for ushering in development, economic growth and mass employment is a myth. Experience and experiments in Latin America, in China and other Asian countries including India has proved this time and again. Still India is pinning hope on this undemocratic, uneconomic, wasteful and environment damaging measure.

Indian SEZs—A Claim

"SEZs are going to play a vital role in the next decade"—said L.B. Singhal, Director General of the Export council for EOUs and SEZs. What does he say—let us note that first. India started SEZ with the EPZ schemes in 1965. The objective of the SEZ scheme is to provide world class infrastructure to manufactures so as to increase exports of goods and services,, provide employment and increase investment from foreign and domestic sources.

EPZs amounted to only Rs. 8000 crore, but crossed Rs. 90,000 crore in the last fiscal. Before the SEZ Act exports were around Rs. 22,000 crore in 2005-06. But in a period of just three years exports grew more than four times. In the past three years (2006-09 there has been an incremental investment of Rs. 1,04,867 crore in SEZs. SEZs at present are providing direct employment to around 3,87,000 people and indirect employment is almost three times of that direct employment in 2007-08; the exports were Rs. 66,000 crore. And in 2008-09 it was estimated to be Rs. 125,000 crore—but the global melt down has dealt a blow to India's exports so that the exports from the SEZs came down to Rs. 90,000 crore this fiscal. This shows clearly India's dependence on Europe and the USA. In the world now SEZs are operating in only 40 countries (out of more than 196). The Indian model on SEZs is quite comprehensive and competitive to any such package provided elsewhere in the world."

The SEZ developers demand more benefits (financial) from the government. The RBI has classified the SEZs as Commercial Real Estate (CRE) projects but they want these projects to be classified as infrastructure projects by the RBI so that SEZs can get funds on the terms and conditions that are available to the infrastructure projects. As per the External Commercial Borrowing (ECB) guideline the SEZs cannot get funds from the ECB—but if those are classified as infrastructure projects they can have access to ECB funding. This demand is simply an attempt to blackmail the government. Because the euphoria which is associated with the operation of SEZs is not going to last long. Allowing them all types of conceivable concession, economic and non-economic even endangering the environment and ecological balance and inviting a rift between SEZ and non-SEZ employees and area, the country is sowing the seeds of social unrest. SEZ economics has failed everywhere—whether in Latin America, North America and Asia—China's experiment with SEZ has failed miserably.

CAG on SEZs: Flaws in SEZ Functioning

The CAG (Comptroller and Auditor General of India) in its limited review of SEZ functioning has found flaws and major weaknesses in their management and functioning. An audit by CAG has shown a revenue loss of Rs. 246.72 crores and additional Rs. 1724.87 crore foregone and non-recoverable in the absence of enabling provisions. It said the sample audit was too narrow to make a general submission. A.K. Banerjee, Deputy CAG said, "In 22 out of 55 SEZs (operational, but notified SEZs are 550) audited, the actual export earning accounted for 28 percent of their total earning, remaining 72 percent coming from domestic sales (which is not permissible).

There was no restriction on "deemed exports" being reckoned

exports enabling the units to attain positive net foreign exchange earning (NEE) predominating through deemed exports rather than actual exports.

The domestic tariff area (DTA) which are outside the SEZ area were at a disadvantage. There is no provision to recover duty foregone on inputs procured by the SEZ units and used in the manufacture of products, which were cleared at 'nil' rate of duty in DTA. This disparity has to be addressed by the government.

This CAG report brings or points out the malpractice in SEZ developers and investors. Dishonesty is their capital of business. Getting unbelievable or unthought of economic advantages and everything at a premium they are duping the country in the name of earning foreign exchange. If this be the picture of less than half of the operational SEZs—then it could be easily surmised what the total show would present.

Revenueloss—for the SEZs

The SEZ developers and the units located there are allowed tax exemption and other benefits and for this the government will loose huge tax revenues. Various projections on estimate of the finance ministry state that the loss may be between Rs. 90,000 crore and Rs. 160,000 crore over the first four to five years of the creation of SEZs. The National Institute of Public Finance and policy said in a report that the special concessions for exports to the SEZ developers and investors are uncalled for in the age of globalization. This amounts to encouraging protectionism in the special economic zones of the country. So the SEZ should be abolished to end discrimination against the non-SEZ area investors.

The SEZs are deemed as industrial townships under article 243 of Indian constitution—which is a dangerous clause because the zones are not under the jurisdiction of municipal corporations, nagar panchayats, village panchayats or any other local authority. The SEZs Act overrides all existing Acts, including that pertaining to local government, so the SEZ Act is undemocratic protectionist, environmentally and ecologically unacceptable.

Even Jagadish Bhagwati, a staunch supporter of globalization, has come out against SEZ. He said, "India does not need SEZs. China needed SEZs because she depended much on export promotion-oriented growth. India is a democratic country and does not grab peasant's land. China does this without any protest (but this is not true because the Chinese farmers are now revolting against the government's SEZ policy and forcible grabing of land), because there is no NGO, no free press, no opposition parties to fight government's policies.

India attempted this very strategy of export promotion with a difference. It was Export Processing Promotion Zone—one such Zone was operational in Falta in West Bengal and some EPZs operating in some other parts of the country—but have not been very much successful. In this state a move was made to build a chemical hub in Nandigram—East Medinipur—covering over 1400 acres fertile land—which would have displaced over 50,000 people. Hundred of schools, temples, mosques, market place, health centre and many other valuable places would have been lost forever. Who will compensate this huge social cost? Most of these people would have turned into simple numbers. Does development mean eviction, displacement of poor people, unemployment and ecological imbalance and overall environmental disaster? Obviously no. In every state whether it is Andhra, Karnataka, Haryana, Orissa everywhere—people are coming forward to resist the move of setting up SEZ in our country,

III. Sustainable Development

From the two previous sections it is seen that the history and experiment with SEZ strategy can help a country temporarily achieve high economic growth rate but not sustainable development. Sustainable growth concept advanced by the Brundtland Land Report was defined as "development that meets the needs of the present without compromising the ability of future generations to meet their needs" (*Brundtland*, 1987). It is "the notion that economic development should proceed at a pace and in a manner that will conserve the environment and depletable natural resources. In its extreme form (steady state growth) human population would be stabilized and renewable resources only would be employed".

Sustainable development is thus heavily dependent on nature. This concept received its first major exposure in 1980 with the appearance of the world conservation strategy, and in 1987 the

World Commission on Environment and Development produced its "Brundtland Report" named after its Committee Chairman the then Prime Minister of Norway. The Report was published under the title "Our Common Future" and from then sustainable development found a place on the United Nations agenda.

Pollution and Development

Sustainable development is closely linked with environment and depletable resources. In a mad rush for 'market economy' or 'globalization' the developed countries forgot to honour nature and environment. More than two hundred years of Industrial development and very recently the theory of development at any cost has made the existence of life (man, animals, plant) in this planet uncertain or precarious. The use of fuels in electricity generation, more and more use of cars, the production of chemicals poison and non-poison, too much greed of men in consumption of meat, the use of fertilizer, insecticide, pesticides and various kinds of chemicals in agriculture have led to global warming. The emission of CO_2, NO_2, NH_4 and CFC and other Green House gases have made our existence here very uncertain. Climate has changed and has been changing very quickly—endangering the existence of this green planet. Ozone layer has been damaged badly to add to man's woes. Disposal of industrial waste is a problem. But the disposal of nuclear waste and electronic waste are much more hazardous problems. We have not yet found out a way for safe disposal or destruction of nuclear waste. These waste materials are very dangerous and deadly poisonous elements. All these pollutants are changing the global atmosphere. Scientists have not yet been able to solve the problem of disposal of dismantled nuclear power plants. These deadly poisonous wastes are heaping up daily alarmingly causing environmental disorder.

Polluted air, water and land and global warming have now appeared to be a hurdle to sustainable development. We the human being often think that the earth is meant solely for us i.e. for human being. This is a wrong notion or idea—because human beings would not be there without the bounties of nature. We cannot live without the help of nature. Whenever men went against nature or worked against nature, she retaliated with much vengenance and force unthought of. India knew this truth from the very beginning of Indian civilization and history. Europe did not accept this truth till the fifties of the last century. They (European) tried to conquer and lord over nature—the result of this is now very clear to all of us. Scientists have, after a thorough search, not been able to locate a single other planet like ours—in the universe—where life can sustain. We need atmosphere which will rear up a living cell for survival.

Europe and America had and still today have a great love for a growth rate which is highly and heavily materialistic and destructive to human well-being both at the physical and psychic levels. The USA attempted for deluding cost of environmental degradation and depreciation of natural capital in the estimation of economic growth in the name of "environmentally adjusted" Real GNP in the 1960s but burned the issue thereafter silently.

Since 1987, after the World Commission on Environment and Development very "little has been done in practical terms to integrate economic development and environment. The Brundtland Report (1987) prescribed a new style of economic development that would sustain human progress in all areas, emphasised that regional and global meeting be held regularly. The United Nations Environment Programme supported this move. The UN General Assembly decided to convene the United Nations Conference on Environment and Development at Rio de Janeiro, Brazil, June 3-14, 1992. It was the most important meeting from the environmental point of view. It was called 'Earth Summit" Conference.

The 'Earth Summit' declaration on environment and development set out 27 principles to guide the international community towards globally sustainable development. The declaration advances the most important economic principles necessary for sustainable development—environmental impact assessment, the precautionary principle, pollutor pays principles, cost internationalization, etc. The conference supported liberalisation of trade as a means of sustainable development. This is nice to suggest—but a fragmented liberalisation/globalization cannot ensure sustainable development; for sustainable development the developed countries should have to come clean—that they really want sustainable development. The developed nations would have to forego a lot of amenities and living standard that they enjoy now. The consumerism 'cult' has to go. The developed countries consume more petroleum products, meat, and many other earthly valuables than the poor third world nations. They let-off more CO_2, CFC and green house gases than the third world countries. They have to agree to reduce emission of all these gases within a specified time, otherwise the goal of attaining sustainable development will remain a distant dream. 'Kyoto Protocol' is a way out of this situation but this protocol has not yet been honoured by the USA.

Population is a very important issue in this new development. It is a fact that world deforestation is going apace. Forests are to be created not destroyed. But to do this we have to stabilise the population growth by 2025. For growing population we need more food, more water, more land for cultivation, more industries to create job opportunity, more schools and colleges for education. So a stable population could help solve these problems.

IV. CONCLUSION

Global environmental deterioration is continuing. It has to be arrested for man's future. Intergenerational equity is to be achieved. This equity has been endangered by environmental disorder. This generation should see or leave its house in order so that "future generations inherit an earth that is at least as environmentally soothing and as livable as it is today".

There is glaring inequality between nations in their use of scarce global resources. Will the developed nations agree to sacrifice a share of the benefits they enjoy for the developing nations? The developed countries have always enjoyed the major share of world's valuable resources. They exploited the poor nations wealth indiscriminately—the question of sustainability was never raised. But when the third world countries are asserting their rights to have equal, nay more share of the "global commons" (ocean, atmosphere) the developed countries are teamed up against the developing countries.

An American on an average consumes 3429 kg petroleum products (oil), a Japanese consumes 2082 kg, a European consumes 1113 kg, an Indian consumes 117 kg only, the world average being 925 kg. Let us look at food consumption by the rich and poor nations. An American used to consume 946 kg of food in 2003 and this amount increased to 1046 kg in 2007. In four years the quantum of food intake increased by 100 kg i.e. 25 kg per year. Whereas an average Indian consumes 178 kg of food and this amount has not increased during the last three years. In America an individual takes 48 kg milk per annum—the amount of milk taken by an Indian is 11 kg. per annum. An Indian consumes 11 kg (per annum) edible oil annually against 41 kg by an American. Indians generally eat less meat but an America eats beef 45-60 kg and chicken 45.6 kg. annually. So the onus of making earth warmer should lie on them and not on the have not nations.

For sustainable development—biodiversity should have to be retained and should not be allowed to be disturbed. Environment and ecology should not be damaged anyway. The earth has not been made the sole preserve of Man alone. Man is a part of the whole ecosystem and environment. It is man who is responsible for environmental disorder and ecological imbalance. It is the greed, too much greed of man which is responsible for the global warming. He is responsible for the big holes in the ozone layer, unplanned and unscientific use of natural and mineral resources, have denuded nature. Population growth all over the third world has, to some extent, led to deforestation, depletion of ground water reserve. Though the truth is that 33 percent people of this world does not get wholesome water. The well-off families waste huge quantum of precious water everyday. Rain water harvesting is an important component of sustainable development.

Development through SEZs is a temporary phenomenon. SEZ's have proved to be a menace to human society and environment: unplanned undemocratic, uneconomic, environment destroying SEZ ones have not only damaged nature in several countries but also exposed man's unsatiable greed and helplessness against natural forces. So, SEZs as we have seen till now should have no place in the map of development for the future generations. Sustainable development is the new *Mantra* for the present and future of mankind, environment, and this unique beautiful planet.

References

Ananda Bazar Patrika, 18.10.2007–1st April, 2007.

Bartaman, 5 May 2008.

Bharat Dogra (2009), "SEZ Act should be withdrawn", *The Statesman*, 23 August.

Bhasker Chakraborty (2008), *Laws Relating to SEZ*, Kanoria Jute and Industries Ltd., Sangrami Sramik Union (in Bengali).

Brundtland, W. (1987), *Our Common Future*, World Commission on Environment and Development, Oxford University Press, Oxford.

Environment Policy and Law (1992), Amsterdam, Vol. 22/3.

Platform, *Bissawan, Silpayan, Unnayan, Chemical Hub*, Kolkata (in Bengali).

Platform, *Chemical Hub: Ekti Nisshabha Ghatak*, Kolkata (in Bengali).

Ron More Jan Ryan (eds.) (1985), *Sustainable Development—Policy and Practice*, New Age Industrial Publishers, New Delhi.

Stiglitz, J.E. (2002), *Globalisation and its Discontents*, Penguin Books, New Delhi.

The Statesman, 12.03.2008.

The Statesman, 13.06.2009.

The Statesman, 21st April, 2007

Sustainable Development and Environment Protection: Role of the Banking Sector

P.S. Tripathi and Shweta Dikshit

During the last half of the 20th century, environment protection and economic development have emerged as objective of every country developed, developing or under-developed. But it has also been a reality that these two issues are often in conflict with each other. This paper tries to highlight the point, that both economic development and environment conservation are twin targets which are complementary to each other and the role of banking sector as a facilitator is of great importance as it is the lifeline of any economic set-up, touching lives of every section of society. It explores the various developments and innovations across the world to achieve this target in this globlised era where economies are inter-linked and role of Indian banks in it.

"Sustainable development," has arisen as new object of development in response to the conflicting scenario. It refers to the maintenance of the process of quality of environmental and social system in pursuit of economic development not only for present generation, but also for the requirements of future generations. Therefore, it encompasses all three aspects of development—economic, social and environmental. It can be best achieved by allowing the markets to work within an appropriate framework of cost efficient regulations and economic agents. Banking sector is the most important economic factor influencing the overall industrial activity and economic growth. In this globalised era, the industries and firms are vulnerable to stringent environmental policies, law suits and consumer boycots. As commercial banks are one of the major stakeholders in the industrial sector, they find themselves more open to credit and liability risks. Any such mis-happening, may turn the 'present performing assets' of the banks to 'future non-performing assets'; thereby affecting the asset quality and profitability in the long-run. Thus banks need to play a pro-active role in making ecological and environmental aspects as a part of their lending principle forcing industries to adhere to the laws, technologies and management system, i.e. 'Green Banking' proving beneficial for industry, banks and economy as a whole. At the international arena, not only governments but also banks are doing much in this regard. Bank track, Equator Principle, Sustainable Banking Award, etc. are few important steps in this direction. However, in India we still are far behind and below the global standards. Though banks have started many schemes to help in the environmental preservation along with economic development; a lot is still desired. Banks play a very crucial role as they deal with large cross-section of clients whose activities affect these issues. It has also been observed that banks sometimes, themselves act as a hurdle in this regard, due to short-term gains whereas as these projects are of long-term gestation and a relatively high cost. Hence, government should provide help to bank in this matter. Bank's role is of paramount importance in this regard, as they being in direct touch can contribute by communicating with every section of the society—literate or illiterate. Thus the role of banking sector is very crucial in attaining the objective of sustainable development; as finance affects economic, social and environment of mankind not only of the present generation but also for future generations.

This paper has been divided into six sections. Section I is Introduction, which explains the meaning and importance of sustainable development and environment conservation, as well as taking into account the role of the banking sector; Section II, titled, "Eco-Friendly Banking and Sustainable Development" discusses the emerging concept of green and eco-friendly banking and its gains; Section III states various developments and efforts which are being made in India and all around the world to balance development and environment. Section IV highlights that sustainable development and environment management are twin targets for which banking sector can play a role of great facilitator. The last section, i.e. Section V brings out the conclusion on the basis of discussion in the proceeding pages and suggestions are made to make further improvement.

I. Introduction

Since the last half of the 20th century environment and development have occupied a prominent position, as we realize day after day that both of these issues are very crucial for mankind and its survival. In this scenario 'Sustainable Development', has emerged as a new object of development in response to this discourse, be it any country—developed, developing or under-developed.

Concern over conflict between environment and development was first addressed at the United Nation Conference on the Human Environment (UNCHE) in 1972 held at Stockholm. Further endorsing this point World Conservation Conference urged conservation as a means of development and utilization of species, natural resources and eco-system. Later many conferences were held world over, stressing on the need of sustainable development.

Sustainable development refers to the process of maintenance of quality of environmental and social system in pursuit of economic development. Renowned economist Robert Repetto has defined, "Sustainable development is a development strategy that manages all assets, natural resources, and human resources as well as financial and physical assets for increasing long-term wealth and well-being."

Thus it can be said that sustainable development as a goal supports those policies and measures which can contribute to raising of living standards, provide education, health, equal opportunities, etc. i.e. all aspects of economic development not only keeping in mind needs of present generation but also of future generations by conserving earth's environment. In this era of globalization, there is an increasing awareness about sustainable development all over the world and joint efforts are being made to ensure that sustainable development is not lost sight of in pursuit of goals such as profit-making, social service, etc. Therefore, it is balancing between development on one hand and environment protection on the other.

In the modern set-up, financial institutions are lifeline of any economic system. Role played by these commercial banks and institutions, as the providers of finance for businesses of all shapes and sizes, have a very pivotal role to play in promoting sustainability across industrial sectors, and communities. In this regard, it should be noticed that sustainability also offer vast potential for financial institutions to improve their own products and services, as new laws on regulations for environment are being enacted by the government. There is no difference between goals of development policy and environment protection as both aim at improving human welfare. Banking sector, especially in Indian context, is not just merely finance provider but also agents of social welfare, thereby, making them most eligible for achieving the goal.

The objective of this paper is to study, how financial institutions and banking sector in particular in this era of globalization and rapid industrial development, can contribute to sustainable development along with environmental conservation.

II. Eco-Friendly Banking and Sustainable Development

Banking sector is generally considered as environmental-friendly in terms of emissions and pollutions. It is this sector which influences the economic growth and development both qualitatively as well as quantitavely, thereby changing the manner of economic growth. How do commercial banks influence the environment in providing finance for development? Environmental impact of banks is not physically related to their banking activities but with the customer's activities, thereby making environmental activity very large though difficult to evaluate as they can be in future as well. These days more and more government across the globe, are taking steps in the direction of environment preservation through legislation, acts as well as increasing public awareness. Therefore, encouraging environmentally responsible investments and prudent lending not only increases the value of the enterprise but it also is a responsibility of banks and other financial institutions.

These days the nomenclature, "Green Banking"; is fast gaining currency. It implies that banks accord priority to those industrial units or projects which adhere to the environmental regulations and also those industries which are helpful in this process of environment restoration. This concept of Green Banking is beneficial to the banks, industry and economy as a whole.

Environment is no longer exclusive responsibility of the government and direct polluter emitters, but also the other partners and shareholders in the business operations like commercial banks, financial institutions which can play a very crucial role in promoting a linkage between economic development and environmental conservation. Role of banking sector is all the more crucial due to their potential reach to the wide variety of the customers. This is applicable also in case of the sustainable development as by helping the lower income group or even people living below poverty line, the commercial banks help to raise the standard of living which in turn affects medical conditions as well as education. By their operations, banks can also propagate importance of sanitation, sewage conditions which pollute rivers (which are major source of drinking water in urban centres), etc. which this section of society can effectively understand through 'oral publicity', as literacy level in low. A study conducted in the United States of America in 2005 by Hamilton and Hart (Blacconiere & Pattern) revealed that environmental performance and financial performance were inter-related. It was found in the study, that those banks which provided funds to those industries which were adhering to the norms of environmental regulation, earned more profits as they were least in danger of closure. Secondly, as people are also conscious that they purchased those environment-friendly products thereby making finance to these units profitable. There have been attempts to adopt sustainable development strategies from various quarters at international level. International consortiums, mutilateral financial and development institutions have been advocating for environmental standards and strategies to evaluate investment projects. In recent years, the International Organisation for Standardization (ISO) has issued a series of detailed outlines for incorporating environmental protection and pollution prevention objectives into industrial activity worldwide, collectively known as ISO1400. These measures would definately provide a much needed impetus for the banking industry to expand the use of environmental information in their credit extension and investment decisions.

III. Development in India and Across the Globe

The financial sector and banking sector in particular, the growing adherence to environment conservation along with a policy of sustained development can be attributed to the recommendations of the UN Secretary General's High Level Panel 2006:

- There is increasing pressure from various agencies both governmental and non-governmental as well as consumers for urgent and effective action.
- Environment conservation and economic development are two sides of the same coin. Economic as well as environmental degradation will have a far-reaching monetary and social implication thereby affecting attainment of Millennium Development Goals.
- In order to reduce the economic gap between rich and poor; and increase the standard of living along with other aspects especially for poor, a multilateral coordinated action programme should be implemented for environmental and sustainable development. More effective implementation of the laws should be strengthened to ensure that desired result is obtained.
- Governments should ensure that policies and laws should be such as they attain a balance between three pillars of sustainable development—economic, social and environmental.

Thus, it is clear that sustainable development is the top priority of which providing sustainable finance is the basis as it effects both society and environment. It is also of paramount importance considering the crucial role of this sector in financing the developmental and economic activities. In this regard, in recent years various laws have been enacted in our country in the areas of air and water pollution control and hazardous waste management. These are very crucial issues for survival of the mankind.

Minimum National Standards for discharge of pollutants in the environment have been drawn up. The Environmental Impact Assessment has been made mandatory for large and medium scale industry. Site inspection report should include review of the past and present use of the site, the nature of the neighbourhood, the production process, status of discharge permits, locations and conditions of storage tanks, etc. The industries which need special attention by banks have been notified by Ministry of Environment and Forest—primary metallurgical industries, producing dyes, paints, fertilizers, pesticide, leather tanning, storage batteries, basic drugs, rayon, sodium/potassium cyanide, acid, plastics, rubber-synthetics, cement, asbestos, fermentation industry and electroplating industry. Banks have been asked to pay special attention to the waste management of these industries as they affect the breathing and rivers thereby influencing the people. Another reason is that if industrial units do not adhere to the norms, they may have to shut down affecting employment and further economic development. Banking sector has been actively engaged in providing assistance to the industrial sector for addressing the environmental issues faced by them. It also makes economic sense as polluting industries may face massive boycotts adding to the cost enormously. So banks are providing scheme for Industrial Pollution Control, Industrial Pollution Prevention Project, the Ozone Depleting substances phase out projects (ODS I, ODS II, ODS III and ODS IV); Industrial Energy Efficiency Project, the Greenhouse Gas Pollution Prevention Project, the Energy Management Consultation and Training Project are important among many. Many banks are now making environmental issues as a part of their internal operations, investment criterion and financial services. In addition to these steps, banks and other financial institutions are creating corporate environmental policies that promote internal energy efficiency and reduce waste. They are also carefully factoring environmental assessments into loan and investment standards.

All these measures are very important in the present context when there is massive concern about environment protection around the world and specially in this recessionary phase, that banks safeguard themselves against conversion of the 'now performing assets' into future 'non-performing assets'. Table 1 reflects the category-wise summary of status of pollution control in seventeen categories of industries in India.

TABLE 1

Category-wise Summary Status of Pollution Control in India (17 Categories of Industries)

Year	*Total*	*Complying**	*Defaulting***	*Closed*
1999	1551	1284	114	153
2000	1551	1326	53	172
2001	1551	1350	24	177
2002	1551	1351	22	178
2003	—	—	—	—
2004	2155	1877	53	225
2005	—	—	—	—
2006	2678	2044	299	335

* Meeting all the standards of Pollution Control.

** Not having adequate pollution control means to comply with norms.

Source: *Report on Pollution Control & Management*, Ministry of Environment and Forest, www.indiastat.com.

Two points can be inferred from the above table—

(1) In India, more and more industries have been adhering to the pollution control norms; and

(2) However, there are equally increasing number of industries which have defaulted or closed down, indicating that due to environment consciousness, laws are more stringent.

It needs to be remembered that banks incur financial losses due to closures as they lead to bad asset and liability.

In the early 1990's the United Nations Environment

Programme (UNEP) launched UNEP Finance Initiative of which some two hundred financial institutions around the world are signatories to promote sustainable development within the framework of market mechanisms towards a common goal of environment. Its objective is to integrate the environmental and social aspect to the financial performance and risk associated with it in the financial sector. Thirty-four international banks follow this conduct. Financial institutions and various governments across the globe are trying to initiate policies for environmental improvements through the development of new financial products and services. In 2002, a global coalition of NGOs in U.S.A., formed a network called, "Bank Tract" to promote sustainable finance in the commercial sector. In October 2002, a group of banks along with International Finance Corporation joined heads to initiate common guidelines for providing sustainable finance known as 'Equator Principles' which revised its guidelines in 2006. These were known as 'Equator Banks'. Presently forty six financial institutions from sixteen countries with business operations in more than ten countries have embraced this principle. IFC along with Financial Times has also initiated. "Sustainable Banking Award" since 2006. It is very important to remember that all these measures aimed at integrating environmental concern into business operations of banking institutions are "voluntary in nature" meant to promote common good of economics and ecosystem for present as well as future generations.

Voluntary commitments have their limitations in this globalized competitive economy and the lenders will always have an incentive to procrastinate their social commitment and prioritize their commercial gains in the short run. Industrial sector has played a very crucial role in India's high trajectory growth story. Our economy is world's sixth largest and one of the fastest growing economies of the world; but it is also second fastest growing country in terms of producing green house gases. Delhi, Mumbai and Chennai are the three of the world's ten most polluted cities. Though government has been trying to address the issue by framing environmental legislations and encouraging environmental technologies, yet the track record is very poor. As such, Indian banking sector and industrial sector face the challenge of controlling the environmental impact of their business, i.e. reducing pollution and emission by their clients. It is therefore, industrial financing by banks assumes a huge significance. As far as international banks are concerned, they are far ahead of Indian banks which are lagging behind the schedules. None of Indian banks have adopted Equator Principle. It is very disappointing to note that in the year 2006, none of the major Indian banks (whether of public or private sector) found a place in the Sustainable Banking Award for leadership and innovation in integrating social, environmental and corporate governance objectives into their operations except Yes Bank (a small player of large Indian Banking Sector) and that too in "Emerging Markets Sustainable Banks" segment.

Thus, it is clearly visible that a lot is still desired by the Indian Banking Sector in this direction. They are still to go a long way.

IV. Environment Management and Sustainable Development by Banking Institutions

The Ottawa Conference IUCN in 1986 outlined five requirements of sustainable development: (1) Integration of conservation and development; (2) Fulfilling basic human requirements; (3) Attainment of equality and social; (4) Right of social self-determination and cultural diversity; and (5) Maintaining ecological balance.

However, there are several hurdles in sustainability of sustainable development, (i) Acceleration in the usage of both renewable and non-renewable resources; (ii) Though the agricultural production has risen but so have desertification, soil erosion and salinization of productive land; (iii) It is estimated that about 40% of the total terrestrial photosynthetic production but energy generation industrial processes, transport, etc. which emit CO_2, SO_2 chlorofluorocarbons (CFCs) and a range of toxic chemicals are putting a much bigger pressure on atmosphere than ever; (iv) One-third of the world's population has inadequate sanitation and more than one billion are without safe drinking water; and (v) The poorer section of the society are exposed to unsafe conditions caused by soot and smoke.

All these activities have resulted in loss of species, biodiversity and damage to the atmosphere. The Brutland Commission rightly pointed out that, "Poverty is both a major cause and effect of global environmental problem." Solving the environmental problems faced by these people will provide impetus in reducing poverty and raising productivity leading to economic development which is sustained equitable and environment-friendly. Hence the need is of integration of economic development, environment conservation and poverty alleviation.

As we have said in the preceding pages that banking sector has a major role to play in this regard. Commercial banks should be concerned with social and environmental issues as through the portfolio of their customers, banks are exposed to these issues than any other average business. In other words, sustainable banking, is the requirement of the times where sustainable development gives advantage and growth opportunity, as it links present with future. The concept of entails three systems: Ecological, Economic, and Social.

All three are inter-linked and banks due to their vast customer base can play a paramount role. In a fast changing economy, where political boundaries are not a hindrance, competition has intensified, the industries and firms are exposed to stringent public policies, tough law suits or boycotts by consumers. These conditions lead to difficulty in recovery of return from the investments for the banking sector. Therefore, they need to play a pro-active role to take environmental and ecological aspect as a part of their lending principle thereby compelling corporate sector to required investment for environmental management, use of appropriate technologies and management system. It is also important because it affects health conditions. Development aims at improving welfare of the people which includes life expectancy, child survival, living conditions, education, etc. along with other broader aspects of economy as wealth creation, consumption, production, etc. Banks are life-line of any economy; therefore they play a pivotal role in achieving twin target of sustainable development and environment conservation. The banking sector in emerging markets is becoming more complex and thus more vulnerable to financial and non-financial risks, as well as increased global competition. The government as well as the central banks are encouraging risk-based approach to lending and investment. They encourage and even provide incentives to incorporate systematic procedures for assessing social and environmental risks. The financial institutions ensure that customers are complying to the prescribed norms but pay less attention after the post transaction period, where the lapse takes place. In India, this practice is very common, but an important aspect. Though, commercial banking pays more attention to the investment banking than environmental problem, yet these are going to play a larger role in the near future regarding investment decisions, otherwise they would prove a hindrance. Schmidheny and Zorraquien (2002), concluded from their primary study that banks can play a hurdling role for sustainable development because:

- They prefer short-term pay back periods whereas sustainable development requires a long-term investment;
- Investment which take into account, the environmental side-effects usually have lower rate of return in short-term. Therefore, sustainable investments are unlikely to find sufficient finding within current financial markets.

In the light of these problems, proper legislation are required for banking sector and as well as ensuring enforcement. In India, banking and financial sector should be made to work for sustainable development as we are very far behind in comparison to international standards.

Sustainable development is development that lasts. A specific concern is that those who enjoy the fruits of economic development today may be making future generations worse-off by excessively degrading the earth's resources and polluting the earth's environment. Therefore, it is more desired that trade-off should be there between development on one hand and environment protection on the other. It implies that the conditions necessary for healthy human life through generations without which development can't be sustainable.

V. Conclusion and Suggestions

Following conclusion have emerged from the discussion on the topic:

1. There is a growing concern all over the world regarding conflict between economic development on one hand and environment conservation on the other.
2. In order to strike a balance between these two issues,

United Nations along with other international consortium has prepared the objective of 'Sustainable Development' implying not only economic but also social and environmental upgradation for present as well as future generations.

3. Banking sector is the major economic agent influencing the economic growth and development both qualitatively as well as quantitatively.
4. It is considered as environmental-friendly in terms of emissions and pollutions where internal economic impact due to use of energy, paper, water, etc. is relatively very low and clear.
5. Environmental impact of banks is huge though difficult to estimate as they influence through the activities of their clients.
6. Green banking opt eco-friendly banking is the order of the day as it is beneficial for bank, industry and economy.
7. Banks are of paramount importance in attainment of the goal of sustainable development as they deal with a large cross-section of society as clients which will have impact on environment.
8. As more and more pressure is being put on the governments of various countries by international consortium and agencies to enact and implement stringent environmental laws, so if banks lend credit to any such industry which does not adhere to norms, may lead to the closure of the unit thereby affecting non-performing assets.
9. The banks due to their personal interaction with the people can contribute more effectively in achieving sustainable development and environment protection as they communicate the importance of the goal, thus helping banks to enhance their reputation as socially responsible organisation.
10. As per the recommendation of the UN Secretary General's High Level Panel, 2006, efforts are being made across the globe, by the governments. Commercial banks in order to felicitate this objective, have formed Bank Track, Equator Principles (which is a set of regulations to strike a trade-off between commercial investment and environment

protection of which many international banks are signatories), as well as many other steps on voluntary basis.

11. In India, efforts are being made in this direction as well various polluting industries have been identified which are to be seen that they stick to the laws of environment protection. Banks have been asked to keep a track of these as well as other industrial units in waste management, etc.
12. Banks on their behalf are encouraging schemes of Industrial Pollution Control, Ozone Depletion Phase Out Projects, etc. They have also made environmental issues as a part of their internal operation, investment criterion and financial services.
13. Though many steps have been taken, but Indian banks are still far behind their international counterparts. In areas of law enforcement and follow-up, our track record is very poor.
14. There is a huge pressure on natural resources due to increasing population and it is not only responsibility of the government but also of financial institutions industrial and corporate sector, common people, i.e. everyone to contribute towards sustainable development in which role of banks is of utmost importance as finance providers.
15. Banks often act as detrimental to the sustainable development due to profit concerns.
16. Banks need to play pro-active role in this regard as sustainable development is one which incorporates economic progress, social upliftment and environment conservation, not only for present generation but also for generations to come.

Following suggestions can further improve the contribution of the banking sector:

1. Government should not just pass a law but also should enforce it strictly.
2. Banks should be given incentives to promote those industries which support environment protection along with employment generation.

3. Banks should be more vigilant in the follow-up of the industrial activities of their clients.
4. They should have power to impose fine on the clients in the form of high interest or any amount if customers are found guilty of lapse in implementation of the laws.
5. Environment impact assessment should be made mandatory in the internal audit along with other operations.
6. In sanctioning leans for housing projects, banks should not only assess the profit but should also ask the builders to develop other aspects like waste management, health centres, etc. In rural areas, Self-Help Groups (SHGs) can be of help.
7. The banks should help in creating consciousness as they are in direct contact with people for whom all the activities of sustained development are aimed.
8. "Think globally, act sectorally" and lend a helping hand to HELP (Healthy Environment and Less Pollution) should be encouraged by the governments and banks.

References

Chopra, Kanchan and Kumar, Pushpam (2005); "Eco-systems and Human well-being; our Human Planet, Summary for Decision makers; *Science & Environmental Fortnightly,* July 31.

Gupta, S. (2003); "Do stock markets penalise Environment Unfriendly Behaviour?" Evidence from India, Delhi School of Economics Working Paper Series; No. 116.

Joshi, B. (2003); *Environmental Movements: A Holistic Perspective,* University Press, New Delhi.

Khanna, Anurag (2006); "Bank Lending and Environmental Protection"; *IBA Bulletin,* March (Sp. Issue).

Munty, M.N. (2007); "India Environmental Outlook"; Discussion Paper Series No. 118/2008; Institute of Economic Growth, New Delhi.

Nayak, Bibhu Prasad and Sahoo Pravakar (2007) "Green Banking in India", Discussion Paper Series No. 125/2008.

Pearce, David and Atkinson Giles (1995); *Measuring Sustainable Development,* Oxford University Press, New Delhi.

Ruya, Yilmaz (2006); "Evaluating Renewabe Energy Sources and Sustainable Development Planning of Turkey"; *Journal of Applied Sciences,* 6(5): 983-87; ISS N1812-5654.

SBI Students Economic Forum (May 27); Theme No. 186; "Sustainable Banking".

Tripathi, Krishnamani and Kumar, Dinesh (2008); "Sustainable Economic Development in the Global Scenarios, *The Journal of Business and Economic Studies*, Vol. II, Issue I.

Magazines /Newspapers

The Hindu.

Times of India.

Economic and Political Weekly (Oct. 2007).

India Today—Various Issues.

The Economic Times.

Website:

www.indiastat.com

Environmental Pollution, Agricultural Production and Livelihood of the Cultivators Around Kolaghat Thermal Power Plant

Subrata Kumar Ray and Sk. Nazrul Hussan

Pollution-free environment is essential for sustaining life of all living beings on the earth. But in recent years the issue of sustainability has become a major concern due to the acute environmental pollution created by rapid industrialisation, urbanisation, deforestation, etc. It is also evident that the demand for energy increased enormously for accelerating the development process. The thermal power is a major source of power generation in a country like India. But the thermal power generation also creates huge environmental pollution through ashes and other poisonous gases discharged by the plant, and this pollution is an obstacle in the way of sustainable agricultural development and good public health. In this context, we have made an attempt to examine the impact of pollution generated by the Kolaghat Thermal Power Plant on agricultural production.

To conduct this study, we have randomly selected sample farmers both from K.T.P.P. command area and non-command area, and made a comparative study (by applying the Dummy Variable Model) to examine the impact of pollution on agricultural production. For this study, we have selected the major crops cultivated in this area, viz. betelvine, paddy and flowers. It is found that the productivity of the selected crops has been adversely affected by the thermal-led pollution. In other words, the productivity of the selected crops is larger in case of K.T.P.P. command area as compared to K.T.P.P. non-command area. This adverse impact on agricultural production is a threat to the livelihood of the farmers living in K.T.P.P. command area. Side-by-side, the pollution produced by the K.T.P.P. creates different types of health hazards and also adversely affects the ecological balance of this area. Thus, the government should take necessary steps like afforestation and proper pollution control system for the sustainable development of K.T.P.P. command area.

INTRODUCTION

Pollution-free environment is essential for sustaining life of all living beings on the earth. But in recent years the issue of sustainability has become a major concern due to the acute environmental pollution created by rapid increase in industrialization, urbanization, deforestation, etc. Vulnerability of the poor indicates exposure to hazards and the livelihood which is below the minimum consumption levels. On the other hand, the opportunities to participate in the socio-economic affairs and interactions in the society have also been deprived of. One of the basic causes of economic backwardness of weaker sections like farmers is the adverse impact of modern production techniques on the environment. It needs necessary development policies to neutralize the adverse impact for accelerating the development of a country like India. Radical changes have to take place in order to make the transition to sustainability within the current century.

It is true that accelerating economic growth and achieving higher standards of living depend upon the availability of adequate and reliable power at an affordable price. Power which is considered to be a pioneer of infrastructure is a part and parcel of human life. It is impossible to think the world without power which is inevitable for economic development. As a key infrastructure component, power increases the production and productivity of agriculture, industry and improves the quality of service sector. In India, agriculture is the backbone of the economy. Creating agricultural surplus in a country like India can accelerate the economic growth. It is to mention that the thermal power is a major source of power generation in a country like India. But the thermal power generation also creates huge environmental pollution through ashes and other poisonous gases discharged by the plant, and this pollution is an obstacle in the way of sustainable agricultural development and good public health in different parts of the country. It is to mention that Kolaghat Thermal Power Plant (K.T.P.P.) situated in a densely populated area of Purba Medinipur district of West Bengal (W.B.) where agriculture plays a pivotal role in the livelihood of the people generates a lot of environmental pollution. But no study has yet been undertaken to measure the impact of this pollution on the agricultural production and livelihood of the workers. In this context, we have made an attempt in this study to examine the impact of pollution generated by K.T.P.P. on agricultural production and livelihood of the cultivators in K.T.P.P. command area.

The objective of the study is to examine the impact of pollution generated by the K.T.P.P. on the agricultural production and livelihood of the cultivators in the K.T.P.P. command area of Purba Medinipur district in W.B.

Data-Base and Methodology of the Study

K.T.P.P. is situated in the Purba Medinipur district of W.B. To examine the impact of the pollution produced by K.T.P.P. we have collected the primary data related to agricultural production, area, net income, etc. from both K.T.P.P. command and non-command area. We have made a regression analysis using the following regression model for a comparative study of the impact of thermal-led pollution on agricultural production and livelihood of the cultivator between the command and non-command area.

$$Y = \alpha + \beta_i X_i + \beta_2 D^* + U$$

where Y = the dependent variable,
X_i = the independent or explanatory variable,
D = the Dummy variable,
* indicates that D = 0 for K.T.P.P. non-command area, and 1 for K.T.P.P. command area.
Other symbols have their usual meanings.

We have assumed that the use of inputs for cultivation remains the same in both K.T.P.P. command and non-command areas. We have also assumed that the soil quality is also same in both the areas. This analysis has been done by considering the selected crops like betelvine, flowers (jasmine and marigold) and boro paddy. It is to note that we have selected those crops which are cultivated in both the areas for making a comparative study. The area, productivity and net income of betelvine, jasmine and marigold have been collected/calculated on monthly basis and that for boro paddy on annual basis. It is to state that the study area is a small farm economy having the dominance of small and marginal farmers. Naturally, the sample farmers have to maximize the revenue from cultivation for the sustainability of their livelihood. Again, the livelihood of the farmers in the study area is highly dependent on the cultivation of cash or commercial crops like betelvine, flowers and paddy. But the cultivation of those crops has been adversely affected due to the deposition of fly ash on the leaves (which disturbs the process of photosynthesis), decrease in soil fertility caused by the ash-sedimentation on the top layer of the soil, irrigated water polluted by ash discharged by the thermal plant, etc. It is also another reason behind the selection of these crops for our study. It is observed that the people of the command area also suffer from various diseases like skin infection, eye diseases, etc. due to the thermal-led pollution. For our study we have purposively chosen fifty households (who depend on the cultivation for their livelihood) from each of the command and non-command areas. The command area belongs to some villages of both Sahid Matangini and Panskura II blocks, and the non-command area belongs to some villages under Panskura II block.

ENVIRONMENTAL POLLUTION, AGRICULTURAL PRODUCTION AND LIVELIHOOD OF THE CULTIVATORS: A MICRO LEVEL STUDY

We have collected the relevant data from fifty cultivators, each from the K.T.P.P. command and non-command area (i.e. the sample size is 100). The objective of this micro-level study is to examine the impact of environmental pollution created by K.T.P.P. on the agricultural production and livelihood of the cultivators. Thus, we have conducted the impact study considering the crops like betelvine, boro paddy and flowers (e.g., jasmine and marigold) cultivated in both the areas.

Let us first discuss the scenario of productivity and net income of the selected crops cultivated by the sample cultivators in both K.T.P.P. command and non-command area. Concentrating on the productivity of selected crops it is observed that the crop productivity is lower in case of K.T.P.P. command area as compared to that of K.T.P.P. non-command area. For example, in case of betelvine 90% of the sample cultivators belong to the productivity (pieces per decimal) ranging from 1001 to 1400 in the K.T.P.P. non-command area, but that ranges from 600 to 1000 in the K.T.P.P. command area. As a whole, it is seen that the average productivity of betelvine is 1200 pieces per decimal in the K.T.P.P. non-command area and is 800 pieces per decimal in K.T.P.P. command area. In case of boro paddy 66% of the sample cultivators belong to the productivity (kg per decimal) ranging from 36 to 40 and 34% of them belongs to the productivity (kg per decimal) ranging from 30 to 35 in the K.T.P.P. non-command area, but in the K.T.P.P. command area 86% of the sample cultivators belong to the productivity (kg per decimal) ranging from 20 to 25 and only 14% of them belongs to the productivity ranging from 26 to 30. Therefore, it is seen that the average productivity of boro paddy is 35 kg per decimal in the K.T.P.P. non-command area and is 25 kg per decimal in K.T.P.P. command area. In case of jasmine 84% of the sample cultivators belong to the productivity (kg per decimal) ranging from 301 to 600 and only 16% of them belongs to the productivity ranging from 150 to 300 in the K.T.P.P. non-command area, but 88% of the sample cultivators belong to the productivity (kg per decimal) ranging from 150 to 450 and only 12% of them belongs to the productivity ranging from 451 to 600 in the K.T.P.P. command area. Therefore, it is seen that the average productivity of jasmine is 450 kgs per decimal in the K.T.P.P. non-command area and is 412 kgs per decimal in K.T.P.P. command area. In case of marigold 84% of the sample cultivators belong to the productivity (kg per decimal) ranging from 300 to 500 and only 16% of them belongs to the productivity ranging from 501 to 600 in the K.T.P.P. non-command area, whereas 90% of the sample cultivators belong to the productivity (kg per decimal) ranging from 300 to 400 and only 10% of them belong to the productivity ranging from 401 to 500 in the K.T.P.P. command area. Thus, the average productivity of marigold is 475 kgs per decimal in the K.T.P.P. non-command area and is 400 kgs per decimal in K.T.P.P. command area. From the above analysis it is seen that the productivity of all sample crops is lower in case of K.T.P.P. command area as compared to the K.T.P.P. non-command area.

Let us highlight the net income earned from the cultivation of the selected crops in the K.T.P.P. command and non-command areas. In case of betelvine it is found that 20% of the sample cultivators belong to the net income (rupees per decimal) ranging from 240 to 300, 56% of them belong to the net income ranging from 301 to 360 and only 24% of them belong to the net income ranging from 361 to 420 in the K.T.P.P. non-command area, whereas 76% of the sample cultivators belong to the net income (rupees per decimal) ranging from 120 to 180 and only 24% of them belong to the net income ranging from 181 to 240 in the K.T.P.P. command area. On an average, it is seen that the average net income of betelvine is Rs. 360.00 per decimal in K.T.P.P. non-command area and is Rs. 180.00 per decimal in K.T.P.P. command area. In case of boro paddy it is found that 80% of the sample cultivators belong to the net income (rupees per decimal) ranging from 351 to 400 and only 20% of them belong to the net income ranging from 300 to 350 in the K.T.P.P. non-command area, but in case of K.T.P.P. command area 76% of the sample cultivators belong to the net income (rupees per decimal) ranging from 251 to 300 and only 24% of them belong to the net income ranging from 200 to 250. Therefore, it is seen that the average net income of boro paddy is Rs. 350.00 per decimal in K.T.P.P. non-command area and is Rs. 250.00 per decimal in K.T.P.P. command area. In case of jasmine it is found that 16% of the sample cultivators belong to the net income (rupees per decimal) ranging from 3300 to 6600 and 84% of them belong to the net income ranging from

6601 to 13,200 in the K.T.P.P. non-command area, but 88% of the sample cultivators belong to the net income (rupees per decimal) ranging from 3300 to 9900 and only 12% of them belong to the net income ranging from 9901 to 13,200 in the K.T.P.P. command area. As a whole, it is seen that the average net income of jasmine is lower in K.T.P.P. command area as compared to K.T.P.P. non-command area. In case of marigold 90% of the sample cultivator belong to the net income (rupees per decimal) ranging from 1200 to 2000 in the K.T.P.P. non-command area, but that ranges from 1200 to 1600 in the K.T.P.P. command area. Therefore, it is seen that the average net income of marigold is Rs. 1600.00 per decimal in K.T.P.P. non-command area and is Rs. 1200.00 per decimal in K.T.P.P. command area. Thus, it may be concluded that the net income earned from the cultivation of different crops is lower in case of K.T.P.P. command area as compared to K.T.P.P. non-command area.

From the above analysis it is seen that the pollution created by K.T.P.P. has made an adverse impact on the crop productivity and net income in the K.T.P.P. command area.

Let us examine this issue with the help of econometric techniques. Let us examine first the impact of environmental pollution created by the K.T.P.P. on the productivity of the selected crops namely betelvine, boro paddy, jasmine and marigold. For this study we have used the following regression model:

$$Y_i = \alpha + \beta_i X_i + \beta_2 D + U$$

where Y_i = the productivity of a particular crop,
X_i = the area under cultivation of a particular crop,
D = the Dummy variable representing the value 0 for K.T.P.P. non-command area and 1 for K.T.P.P. command area.
Other symbols have their usual meanings.

In case of betelvine cultivation, the estimated regression equation is:

$$Y_{1i} = 11.272 - \underset{(-2.949)}{0.0458^*} X_{1i} + \underset{(16.274)}{16.006^*} D \qquad R^2 = 0.822$$

where Y_{1i} = the productivity of betelvine,
X_{1i} = the area under betelvine cultivation,

and * implies that the coefficients are significant at 5% level. The estimated regression results show that the environmental pollution has created an adverse impact on the productivity of betelvine in the K.T.P.P. command area.

In case of boro paddy cultivation, the estimated regression equation is

$$Y_{2i} = 11.807 - 0.00556\ X_{2i} + 10.384^{*}\ D \qquad R^2 = 0.953$$
$$(-1.335) \qquad (34.876)$$

where Y_{2i} = the productivity of boro paddy,
X_{2i} = the area under boro paddy cultivation,

and * implies that the coefficient is significant at 5% level. The estimated regression results show that the environmental pollution has created an adverse impact on the productivity of boro paddy in the K.T.P.P. command area.

In case of jasmine cultivation, the estimated regression equation is:

$$Y_{3i} = 59.214 - 0.671\ X_{3i} + 12.667^{*}\ D \qquad R^2 = 0.486$$
$$(-3.656) \qquad (7.104)$$

where Y_{3i} = the productivity of jasmine,
X_{3i} = the area under jasmine cultivation,

and * implies that the coefficient is significant at 5% level. The estimated regression results show that the environmental pollution has created an adverse impact on the productivity of jasmine in the K.T.P.P. command area.

In case of marigold cultivation, the estimated regression equation is:

$$Y_{4i} = 5226.476 - 5.584\ X_{4i} + 18655.416^{*}\ D \qquad R^2 = 0.881$$
$$(-0.087) \qquad (16.203)$$

where Y_{4i} = the productivity of marigold,
X_{4i} = the area under marigold cultivation,

and * implies that the coefficient is significant at 5% level. The estimated regression results show that the environmental pollution has created an adverse impact on the productivity of marigold in the K.T.P.P. command area.

From the above regression analysis it is evident that the environmental pollution created by the K.T.P.P. has made an adverse impact on the productivity of the selected crops. This adverse impact may affect the livelihood of the cultivators. Under these circumstances, we have made an attempt to examine the impact of differential productivity created by the pollution generated by K.T.P.P. on the total net income (adding the net income earned from selected crops) which plays a significant role on the standard of living or livelihood of the cultivators. For this study we have used the following regression model:

$$Y_i' = \alpha + \beta_i X_i' + \beta_2 D + U$$

where Y_i' = Average net income per decimal (from all selected crops),
X_i' = Average productivity (from all selected crops),
D = the Dummy variable representing the value 0 for K.T.P.P. non-command area and 1 for K.T.P.P. command area.

Other symbols have their usual meanings.

The estimated regression equation is:

$$Y_i' = 203.00 - \underset{(-4.48)}{0.396^*} X_i' + \underset{(5.222)}{78.347^*} D \qquad R^2 = 0.355$$

The above results show that the environmental pollution generated by K.T.P.P. has also made an adverse impact on the income. Thus, the environmental pollution produced by the K.T.P.P. has made an adverse impact on the agricultural production and it has become a threat to the sustainability of the agricultural growth in the K.T.P.P. command area. Besides that, pollution also generates impediments in the way of improvement of the standard of living which is a threat to the present and future generation of the sample cultivators. Side by side, it is also found that the thermal-led pollution has also created different health hazards like skin diseases, asthma, etc. which are also major concerns threatening the sustainable development of the K.T.P.P. command area.

CONCLUSION

The availability of adequate and reliable power at an affordable price is an urgent need for economic development of a country like India. But the generation of power like thermal power has many bad consequences which may endanger the environment. From the above analysis it is found that the pollution created by the K.T.P.P. has threatened the sustainability of agriculture and also diminished the living standard of the small and marginal farmers. Since, the K.T.P.P. authorities have not taken appropriate steps to reduce the pollution up to the permissible limit, so the government should implement appropriate anti-pollution policies for sustainability of agriculture and better well-being of the poor cultivators in the K.T.P.P. command area.

REFERENCE

Gujrati, D.N., 'Basic Econometrics', Tata McGraw Hill.

Pollution Impacts of the Indian Sponge Iron Industry

Rifat Mumtaz

India's sponge iron industry has sustained secondary steel producers who use electric arc furnaces or induction furnaces to make steel. Sponge iron has in fact become a perfect substitute for scrap, the availability and price of which have greatly hampered steel production. Post-2001, India has emerged as the world's largest producer of sponge iron, accounting for 20% of global output. There are over 400 plants operating in the country. In 2007-08, India produced close to 20 mt of sponge iron.

The cost of this spectacular growth, however, is being borne by people living in the states that produce sponge iron, as sponge iron units are responsible for unprecedented levels of pollution. Thick black smoke, contaminated water, depleting vegetation, falling agricultural yields, premature death of domestic cattle, and poor human health conditions are just some of the impacts. The plants are located deep inside forested regions that are rich in iron ore and have already been devastated by the mining industry. They include the states of

> Orissa, Chhattisgarh, Jharkhand, West Bengal, Goa, Maharashtra, Karnataka and Andhra Pradesh.
>
> This polluting industry has been classified as the 'Red Category' industry by the Ministry of Environment and Forest and is the most evident example of poor environmental governance that a country could boast of. The Sponge Iron industry, with its clear backward and forward linkages of equally polluting activities—iron ore and coal mining on one hand and steel production on the other is a classic example of how the poor continue to bear the brunt of large scale pollution of their living environment while the polluters continue to flourish with abetment from every, virtually dysfunctional, government or regulatory authority in this country.
>
> Lack of political will and absence of corporate accountability dominate the scene. Economic viability of industry rules over viability of poor lives. Neither the government nor the corporates care about where the plants are sited, how they operate and who is bearing the costs. This apathy has upshot the spontaneous movements that have emerged in the scenario of extreme pollution and inaction by the official authorities. In far off villages away from the centers of popular debates on climate change and global warming the spontaneous voices from these movements have been successful in throwing light in the face of the deficient environmental governance in this country.

In the Indian context, where majority of the poor base their life and livelihoods on natural resource, environment is an issue crucial to the lives of the people at the grass-roots. When environment is hurt, it is they who suffer. Environmental degradation has been playing havoc with the lives of the poor people. Policies at the national as well as the State-level undermine dependence of poor communities on natural resources and tend to look at it from the perspective of commercial commodity or pure conservation.

- Rivers have been sold to industries by negating community access and user rights.
- Ground water sources have been highly exploited by the industries with government permission.

- The coasts have been most often used to dispose sludge and dumping wastes and waste water from industries.
- Forest land and forest resources have been exploited be it for large dams, mining, highways and industries.
- Industries at the heartland of rural India, Adivasi belts and at coasts, have been allowed with a thought among policy-makers that there is no harm in polluting these distant habitats and that the poor communities, already pushed to the margin of any development, have better capacity of absorbing toxics and pollution.

On one hand, communities inhabiting the natural resources hub areas, be it anywhere in the country, have been bearing the cost of the mad race for industrialisation without ever being given any space in decision-making or share in the benefits accrued, while on the other hand, environmental sustainability has been rarely considered as a priority in the Indian policy-making.

The scenario of environmental degradation and commercial exploitation of natural resources has intensified in the era of globalization and liberalization. Policies governing industrialization and related environmental impacts have gradually weakened. Privatization and disinvestment in industrial sectors have hit a high with more private players investing solely for profit-making. The gravest industrial disaster of the country, the 'Bhopal Gas Tragedy' which shocked the world could not awake the Indian Government to ever consider issues of environmental governance seriously.

Environmental Governance envisages a governance structure with pro-people conservation policies to balance justiciable equity in environmental burden. It has the purpose of attaining environmentally-sustainable development. However, this governance mechanism, open to corruption and political influences, in a country like India, has been diluted over past some years and failed to touch upon equity in environmental benefits among various stakeholders. It is in this context that the growth of sponge iron industries in India needs to be critically examined and questioned.

In this paper I have attempted to do an issue analysis of the Sponge Iron Industries and their Environmental Impacts. The paper is based on factual references from the states where this industry is operating under almost a non-functional environmental governance structure.

THE SPONGE IRON INDUSTRY

Sponge Iron or DRI (Direct Reduced Iron) is a material obtained from the direct reduction of iron ore and is used in the production of steel in its various forms. The growth in infrastructural and industrial construction has created a huge demand for steel in the global market and in turn a high demand of sponge iron leading to a rapid growth of the sponge iron industry in the past two decades. Sponge Iron has evolved as the best and cheap metallic substitute in steel making replacing imported scrap, the traditional feed in steel production.

Till the 90's there were only three SIUs (Sponge Iron Units) in the country and these were public sector undertakings. It was the introduction of the DRI technology and use of coal in sponge iron-making, which made sponge iron production a cheap, low cost investment affair that had an open domestic steel market as its consumer.

To facilitate growth of the steel industry, in 1985 the government has de-licensed the iron and steel industry and removed it from the list of industries reserved for the public sector. Over the years, pricing and distribution of steel have also been deregulated. To support availability of adequate finance, 100 per cent Foreign Direct Investment is permitted in this sector under the automatic route. Customs duty on raw materials has been progressively reduced to meet the requirement of the industry and to further reduce the cost of production of steel, liberalisation of imported technology was promoted.

The biggest support to the sector is the government assistance in obtaining non-coking coal, allocation of captive coal blocks, iron ore, natural gas and above all easy land acquisition.

According to the Ministry of Steel, in 2007-08, the installed capacity of Sponge Iron in India reached a whopping 20 million tons from 12 lakh tons in 1990-91. The market projection estimates show that in the next five years, another estimated 70 lakh tons installed capacity of sponge iron will be created due to huge investments in expansion of existing projects and upcoming new

Greenfield Steel projects. There is no scope for exporting Sponge Iron. But the growth in the export of steel is definitely pushing up the demand for Sponge iron in India. The National Steel Policy 2005 also envisages sponge iron production to grow at a rate of 6.5% per annum, with production capacity at 380 lakh TPA (Tonnes Per Annum) by the year 2020.

Growing Global Steel Market and India

In 2006, among global steel exporters, India ranked 11th from its previous position of 12th. According to the estimates of the International Iron and Steel Institute (IISI), till 2010 global demand for steel will grow by 4.9 per cent, but India will see a much faster growth in consumption of 7 per cent. Multinational companies are now setting up manufacturing facilities in emerging economies like India because it has abundant natural resources, both in terms of raw material and energy and a much cost-effective option for these industries.

THE PLAYERS IN SPONGE IRON INDUSTRY

The first sponge iron plant in the country was set-up at Paloncha in Andhra Pradesh with a capacity of 3 lakh TPA in 1980 by a public sector company Sponge Iron India Limited. Post the deliscencing and more so in the 90s, with growing subsidies to the private sector, the heightened growth of sponge iron industry in India was contributed by big private players such as TATA, Jindal, Bhushan, Monnet and Rungta. The mineral rich regions of the country, mostly the Central Eastern belt, became a hub for the growth of mini SIUs consisting mainly of 50 and 100 TPD kilns owned by small scale businessmen and entrepreneurs. Among the coal-based producers are the Jindal Steel & Power, Prakash Industries, Tata Sponge Iron who occupy the top three positions and Essar Steel, Ispat Industries and Vikram Ispat are the top three gas-based producers.

Gujarat and Chhattisgarh are Highest Producers

In 2006, among the states producing DRI, the state of Gujarat emerged as the largest producer of sponge iron in the country. For April-July, Gujarat produced 11 lakh tonnes with Chhattisgarh producing 10.84 lakh tonnes. These two states accounted for nearly 62% of India's total sponge iron production. Maharashtra followed as the third largest producer in the same period with a 19% share and Orissa with a 9% share in the total sponge iron production. Collectively put together, these four states control 90% of the country's total sponge iron production. *(Source: Gujarat Industrial Commissionerate)*

EMERGING ISSUES WITH THE GROWTH OF THE SPONGE IRON SECTOR

The DRI industry is of more than Rs. 6000 crore worth of investment and is flourishing under the existing post-liberalization industrial policy regime where entrepreneurs are free to set up SIUs anywhere in the country based on their commercial judgement.

While gas-based sponge iron production is difficult for most entrepreneurs due to scarce availability of natural gas and its capital intensive nature, coal-based units are relatively easy to set-up as they can be established with the help of local fabricators and suppliers. A 100 tonne per day (TPD) sponge iron plant with the initial investment of Rs. 12-16 crore can make a profit of about Rs. 60 lakh a month within a year's time. *(Down to Earth, 2006).*

And it is this factor which has given impetus to an unregulated and geographically concentrated growth of the small scale SIUs determined only by easy and cheap availability of iron ore, coal, power, cheap labour, a growing and easily accessible market for their produce (rolling mills and steel units), and a state government which is willing to subsidise costs of all required inputs.

As a result, even fly by night companies have got into making sponge iron so much so that many of them are operating without obtaining the mandatory consent from the state authorities which actually makes it impossible to ascertain the total number of SIUs operating in the country. *(http://www.steelworld.com/a%20nerve.htm)*

These units do not have captive mines and power plants as do the big companies. They buy their iron ore and coal from the local open markets through auctioning or tie up with the local-mining mafia.

Some of the issues of prime concern with the unregulated growth of this industry are:

- the area where these factories are mushrooming are the richest natural resource pools with the poorest populations belonging to the adivasi and dalit communities;
- these are also areas where other iron ore, coal mining and steel mills are concentrated;
- the process of coal-based sponge iron making is highly polluting and a presence of these industries in a concentrated form has very hazardous impacts on the environment and livelihoods of the area where these plants come up.

According to the 2007 annual report of Sponge Iron Manufacturers Association (SIMA) "most such (small) units have not installed adequate pollution control equipments such as the ESPs (Electro Static Precipitation Device) and the government authorities have not been sufficiently vigilant on this count. These mini SIUs have been polluting continuously and heavily for the past five years. The petty businessmen resort to taking shelter from the courts under the guise that stopping their running plants would involve laying-off a productive workforce and after obtaining a "stay order" they continue to pollute the environment."

While this is what the SIMA claims, experience from the ground indicates that the polluters and violators of norms are not confined just to the smaller producers. The best way of lowering production costs for all SIUs has been ignoring occupational safety and pollution control regulations and relying on a contractual labour force. The largest companies in this business have indulged in various tactics to gain easy and speedy environment clearances, cover up their mistakes and maximise profits by cutting on costs.

POLLUTION FROM SPONGE IRON INDUSTRIES

A visit to an average Sponge Iron 'ilaka' or an area dominated by these factories leads to some immediately glaring revelations—the air is laden with smoke, the movements of trucks is high, the plants are located adjacent to or right in the middle of a village or villages, the houses of these villages—roofs and walls—are covered with dust, the leaves and forests in the vicinity are black and not green in colour, layers of soot accumulate on the skin, the eyes experience a burning sensation if long hours are spent in the area, dumps of char and iron ore scrap lie along the roadside.

As per a report in *Down to Earth*—"A typical 100 TPD sponge iron plant consumes 160-175 tones of iron ore, 120-150 tones of coal, 3.5-5 tones of dolomite and 120-160 tones of water. In return they emit 180-200 tones of carbon dioxide, 1-2 tones of dust (if they are equipped with required pollution control equipment, otherwise dust emissions can be as high as 10 tones per day), 25-30 tones of char, 10-15 tones of dust from pollution control equipment and 2.5-3.5 tones of kiln accretion every day. The total pollution load from a 100 TPD plant can be as high as 230-250 tones per day."

The type of pollution caused by the emissions and wastes of SIUs can broadly be classified into three parts—air, water and soil (or the land). While the three are inter-connected and the contaminants remain more or less the same, we divide them in order to understand the different natures of the impacts that these have through different sources.

Air Pollution—SIUs emit what is referred to as 'suspended particulate matter' or SPM which contains cadmium, nickel, hexavalent chromium (most dangerous through air and water), arsenic, manganese, and copper which are considered fatal even in small doses. These carcinogenic wastes are emitted from the stacks or the chimneys in the plant. Lower the height of the stack, the more the probability of the emissions settling in and around areas closer to the factory.

While this is a direct form of air pollution that occurs from the stack most of the sponge iron hubs and areas are facing issues of widespread vehicular pollution by trucks that carry the raw and finished material. The spread of dust from loading and unloading of raw materials is yet another hazard.

And solid waste generated by these plants in form of char and fly ash which is openly dumped in or outside the factory premises in fields and forests is carried by the wind polluting the air further.

Water Pollution—Many a times the solid waste is dug into the ground, polluting ground water. The direct disposal of industrial effluents and coal washeries into rivers and streams is a common phenomenon. The other method that is used for disposal is building of waste water ponds in the factory campus. This stored toxic water then seeps into the ground contaminating the ground water. Even the largest sponge iron producer company like Jindal Steel and Power Ltd. has a practice of dumping fly ash in the nearby school ground, river sides, forest areas and road sides.

Industrial dust which settles on every object/tree/plants, etc. within several kilometers from the plants is washed by rain water in the monsoons and drains into the earth through water bodies and the soil.

Ispat Industries operating in villages of Dolvi, Raigad, Maharashtra, present a unique case of coastal impacts due to sponge iron industry. Due to the heavy mobility of vessels and effluents from the integrated steel plant released into the creek, water pollution has increased manifold and it has badly impacted the growth of mangroves and fish availability in the creek. The area comes under CRZ—1 and 2 area under the Coastal Regulation Zone Notification, 1991, as per which no development activity can take place with a stipulated shore line. However, in Dolvi, Ispat is operating for the last 12 years in violation of the CRZ notification and other environmental legislations with the support of state government.

Soil Pollution—The processes of both air and water pollution, directly or indirectly, affect the soil around the plant area as rain water with dust laden top soil percolates into the ground. The disposal of industrial wastes directly in open spaces affects the soil and its fertility.

IMPACTS OF POLLUTION

1. On Human Health

Two preliminary risk analysis assessments done by an environmental engineer, Sagar Dhara (from Cerena Foundation, Hyderabad) in Mayurbhanj (Orissa) and Raigarh (Chhattisgarh) have indicated that the populations in the periphery of the Sponge Iron industries are at a high risk of cancer. "Iron acts along with other carcinogenic heavy metals to increase cancer risk. The toxic effects of heavy metals are varied and may often manifest after a prolonged period, sometimes several years, as in the case of cancer. SIUs also emit oxides of sulphur and nitrogen and hydrocarbons. These air pollutants are likely to increase the incidence of respiratory tract ailments, e.g., cough, phlegm, chronic bronchitis and also exacerbate asthmatic conditions" says the study.

A study conducted by the Chttiranjan National Cancer Institute has shown that almost all school children in the Durgapur, Asansol belt suffer from upper and lower respiratory tract infection. (Source: Industrial Pollution in Durgapur, Nagrik Mancha, Feb. 2005)

The other health impacts include:

- Damage to the nervous system especially among children due to exposure to lead and mercury.
- Dangers of kidney ingestion due to mercury.
- Skin irritation and various other skin diseases.
- Impact on Women's health—reproductive systems.

Dr. Arvind Patil, General Physician based in Bellary, Karnataka, mentioned in an interview that the condition by the KSPCB to maintain minimum 5 rows of trees inside and outside of the unit's boundary wall is never followed seriously by the sponge iron industries. He explained that the health condition of workers working in these industries is pathetic; their lungs are at high risk and in long run it will have worse impact on their life expectancy. (Source: infochangeindia.org/200812267552/Environment/Features/Rifat Mumtaz/Dark-clouds-over-India 's-sponge-iron-industry. html)

Increasing accidents due to heavy traffic and movement of trucks in areas like Keonjhar and Raigarh is another common occurrence leading to deaths and other physical disabilities.

2. On Health of Livestock

It is not just human health that is affected by the pollution. Domestic animals, especially the livestock cattle are being affected by the pollution. The crop residue and grasses that the cattle feed on, the air that they breathe and the water they drink adversely affect their health. The milk that they produce and meat also is contaminated with toxics which in turn affects the human beings on consumption.

Villagers often reported that "the death of these cattle is very painful because of the pollution from the sponge iron industries. Other domestic animals like dogs have also been affected by the pollution."

The farmers from Pinjarala, Kotturu in Mahabubnagar district in Andhra Pradesh reported about the SIUs dumping its waste in open spaces within the village. "Every 2-4 months there is a showdown in front of the factory gate. We have gone with carcasses of our cattle and had several protests. But our village leaders and influential people in the village have come under company influences." Besides, the company takes help of the police to suppress the local protests. They also mentioned that it was common for officials to visit the area for inspecting the plant. However, so far they are yet to get any relief. (Source: Infopack on Sponge Iron Industries: Issues and Campaigns, NCAS, 2007)

3. On Agriculture and Crop Production

Villages in the peripheries of SIUs have their agricultural lands rendered almost unproductive with the increasing accumulation of dust and air emissions on the soil. The quality of the produce has also been affected. "While cooking the rice we can see the amount of dust in the grain as it collects in the vessel" villagers from sponge iron industry areas complain. "The crops and fruits are damaged and yields reduced even at low concentrations of oxides of sulphur and nitrogen."

"In Bromorposhi, a village that lies close to the Shiv Shakti Sponge Iron in Mayurbhanj District of Orissa, it was estimated that the annual revenue loss to the village because of lowered paddy and fruit yields was Rs. 18 lakhs per annum in 2001-02 *(Source: Risk analysis study, impacts of sponge iron industry, Mayurbhanj, Dhara, 2002)*. It is probable that this has been caused by the sponge iron plant's air emissions. It is probable that the SIUs in Raigarh District in Chhattisgarh, which has more than 30 sponge iron units operating, may cause a similar order of revenue loss due to decreased crop yields," says the study by Sagar Dhara.

Impacts of SIUs in Siltara, Raipur, Chhattisgath

As per news reports, in Simga, Siltara and Mandir Hasaud in Raipur district (Chhattisgarh) around 50,000 acres of land has become barren due to heavy industrial pollution in the area and impacted 1,25,000 farmers. A study conducted by Raipur-based Indira Gandhi Agriculture University in 2002-03, indicated that the complete area in this belt will become infertile in the next ten years if industrial pollution continues. Notably, in this area, almost 1 lakh acres of land was under cultivation of which 50% has already become infertile. According to the report, agricultural land within 2 kms of industries is worst affected due to black dust, smoke and particles of carbon and heavy metal released. Sulphur Dioxide, released from the industries, converts into Sulphuric Acid comes in contact with water polluting ground and other water bodies in the area. It was also found that dermatological diseases and Asthma has increased in the area due to heavy industrial pollution. (*Source*: *Chhattisgarh mein pradoshan ke khilaf sangharsh*, NGM, 2005.)

4. On Forest Biodiversity

While transfer of forest and village commons towards setting up of the SIUs and related operations has led to direct deforestation, the impact of pollution can also be seen on flora and fauna. Trees like Mahua, Lac and Tendu are being easily affected by the dust fall. This directly impacts the livelihoods of people as these trees contribute to income for the forest dependent communities.

Decreasing Tendu Leaf Collection from Co-operative Societies of villages located near Sponge Iron Plants in Raigarh District, Chhattisgarh, 2002-05.

	Patrapalli Cooperative Society— *Patrapalli, Chiraipani, Khairpur, Kalmi, Bhagwanpur, Bhelwatikra, Uchbitthi, Dewanmoda, Amapali*	***Taraimal Cooperative Society—*** *Taraimal, Ujjalpur, Jinghol, Punjipatra, Tumidih*	***Tilga Co-operative Society—*** *Karichapar, Tilga, Bhagoda, Kotmar*
2002	649.345	1188.555	832.9
2003	681.85	1210.8	805.35
2004	709.265	1465.8	707.11
2005	284.415	501.5	634.55

Source: Infopack on Sponge Iron Industries: Issues and Campaigns, NCAS, 2007.

The black dust on the forest floor and leaves affects regeneration as well as health of the wild herbivores dependent on these for food. Pankaj Oudhia, an agricultural scientist based in Raipur, Chhattisgarh, who has also conducted a rapid assessment at

Raigarh, found that "... *there is a reduction in biomass, canopy size of the trees and a subsequent reduction in leaf size as a result of the air pollution*".

An interesting example is of the impact on Kosa silk production in Raigarh due to change in the leaf quality of the Arjun tree on which the silk worm feeds.

It is important to note most of these units are located in areas where—

- There are more than one SIUs located.
- Iron ore and coal mines are close by or accessible—for access to the raw material.
- Iron crushers and coal washeries operate.
- Rolling mills and other Induction Furnaces are located where sponge iron can be supplied.
- There is regular movement of heavy vehicles for transportation.

And therefore the cumulative impact of pollution by all these activities is much higher and graver.

EXISTING LEGAL/ENVIRONMENTAL PROVISIONS FOR SIUS

The three key legislations that apply to this industry in the context of pollution are the Environment Protection Act, 1986, the Air and Water (Prevention of Pollution) Acts, 1981 and 1974. The latter confers the responsibility of monitoring, control and prevention of Air and Water Pollution on the Pollution Control Boards.

The Sponge Iron Industry is covered under the EIA Notification, 2006. As per this, the setting up of a Sponge Iron Plant of a capacity more than 200 TPD would require an environmental clearance under category A from the Central Government's Ministry of Environment and Forests. All SIUs under 200 TPD capacities would require a clearance at the State-level. The public hearing or consultation process would only be required for category A projects. All category B projects would be exempt from the public hearing process. Since most of the plants are under 200 TPD, the people will never get an opportunity to voice their opinion on the matter, before a unit comes up.

Under the Central Action Plan, the Ministry of Environment and Forests has classified all hazardous industries into three types—Red (high pollution potential), Orange (medium pollution potential) and Green (low/insignificant pollution potential) for the purpose of surveillance and monitoring. All processes involving the making of steel and the use of a blast furnace or coke oven would fall under the RED category—including Sponge Iron manufacturing. However, it is essentially up to the State Pollution Control Boards to implement this classification and monitoring provisions for it. There is no centrally enforced rule or guideline around this.

In 2005, the Central Pollution Control Board issued a Draft Standards for Prevention of Pollution in SIUs on its website. This process was undertaken after some of the State Boards received regular complaints about the functioning of these plants. Further, earlier in the same year, the AP Pollution Control Board had introduced similar siting guidelines for SIUs. It was in 2008, after delay of 3 years that the Central Guidelines and Emission Standards for Pollution Prevention from sponge iron units have been approved by the Central Ministry of Environment and Forest.

However, these approved Guidelines and Emission Standards for Pollution Prevention from sponge iron units do not have the requisite standards which were proposed earlier. Rather they are more lenient and diluted to the extent that now it facilitates the growth of this industry without considering environmental pollution. Some features of these guidelines are:

- It has raised permissible levels for fugitive emissions from 1,000 microgram/cubic meter to 3,000 microgram/cubic meter for existing units and to 2,000 microgram/cubic meter for new units.
- As against draft standards which suggested 75 m stack height, the final guidelines have brought it down to 30 m.
- It failed to mention about solid waste disposal standards. Initial draft had included char, kiln depositions, scrubber sludge and flue dust as solid waste and had suggested recycling measures.
- Does not say anything about noise pollution.

- No guidelines for industry siting specified. It was well suggested in the initial draft.

(Source: DTE, Juneja S, Slack Approach, August 16-31)

These remain inadequate in the face of the current problems created by the industry. This in a way, opened easy doors for many more plants to establish and to further expand the existing ones.

NO IMPLEMENTATION OF POLLUTION CONTROL MEASURES

The small scale units have been in the line of fire as they are the ones unaware of most of the pollution control norms and related legislations. There have been examples in states like Orissa and Chhattisgarh where plants have been functioning without consent from the Pollution Control Boards. Other norms that are flouted by these plants include,

- Non-compliance to consent and clearance conditions.
- Violations as far as disposal of solid wastes is concerned.
- Violations of Supreme Court Guidelines on hazardous wastes—no boards are put up as per the statutory provisions describing the activity within the unit, etc.

Most of these units do not have the capacity to deal with hazardous processes and wastes. The pollution control measures which includes using of Electro Static Precipitators on the stacks is supposed to help reduce the emissions but most of the small units do not install these and those who have, do not run them as the costs of running the ESP, being a power consuming device is high.

As per Assistant Environmental Officer of Karnataka State Pollution Control Board (KSPCB) based at the District Pollution Control Board (DPCB) office in Bellary, around a year and a half ago, none of the sponge iron units had employed pollution control equipments like the ESP. Persuasion by the Board resulted in convincing only 4 units operating in the Hospet Road area to run the ESP more regularly than others. The KSPCB has not put a compulsion on the units to install an interlocking system between the power supply and the kiln. This was often suggested by the experts. The mechanism will ensure regular running of pollution control equipments because it automatically disconnects power supply to the kiln if the ESP is off. (Source: infochangeindia.org/200812267552/ Environment/Features/Rifat Mumtaz/Dark-clouds-over-India 's-sponge-iron-industry.html)

While the larger plants are in tune with the procedures, they too have been flouting many of the rules. They mostly operate on muscle power and have enough clout within the state and Central government to cover up their errors and violations (e.g., Jindal Steel Plant (JSPL), Raigarh, Chhattisgarh).

- **Mining beyond Permissible limit**—In 1997 JSPL got an environmental clearance from the MoEF and consent to operate from MPSPCB to excavate 2 million ton coal per annum from Dongamahua (705 hectares). In 2006, the mining department found out that JSPL had excavated 60,64,402 ton coal more than the permitted limit of 13,50,000 metric ton for 5 years, violating the approved mining plan of the company. A case has been filed against the company by the Chhattisgarh PCB in Raigarh session court under violations of sections 37, 40 of Air Pollution (Prevention and Control) Act, 1981, sections 44, 47 of Water Pollution (Prevention and Control) Act, 1974, sections 15, 16 of Environment Protection Act, 1986 and section 34 of Indian Penal Code. Mineral Department has also issued notice under revenue pilferage and fined company to pay approximately six crores including fine and stamp duty and registration fees. Instead of adhering to the notices, JSPL ignored the matter and did not pay any heed.[1]
- **Illegal Land Acquisition**—With the help of the district administration, land Acquisition in Raigarh district done by JSPL for its sponge iron plant has been equally unlawful. Poor villagers have not been paid their share of compensation promised by the industry. Since 1992 till date, people are running to the district offices demanding compensation. Today, villages such as Patrapali, Saraipali, Kalmi and many exist in the district map but have been merged into JSPL and became history in reality.
- **Illegal Sale of Electricity**—The JSPL case of violation of the Electricity Act, 2003 is another gross defiance which has been registered with the Chhattisgarh State Electricity

Regulatory Commission. In the year 2005, the Chhattisgarh State Electricity Board (CSEB), filed a petition to prevent JSPL from supplying electricity to industries in its industrial park on the ground that the petitioner neither had a transmission nor a distribution licence and hence supply of electricity by JSPL to industries was without legal sanction under the Electricity Act. The industry was already supplying electricity to 18 industries in its industrial park located about 25 kms away from Raigarh town without securing a prior permission from the regulatory commission as stipulated under the State Government no objection given in 2003 on the condition that the JSPL will obtain licence from the State Electricity Regulatory Commission (SERC) as soon as it is set-up. Even after formation of the SERC within a year the JSPL did not approach the SERC for a licence but continued supply of electricity to the industrial park units and applied to SERC only in 2005.[2]

- **Power play**—In past years there have been regular instances of harassment and village roadblock by JSPL to prevent villagers to enter. The story of killing of a tribal youth named Sitar, who was opposing encroachment of their grave land in Saraipali, in 2006 by JSPL goons is still fresh in the minds of locals, living under a threat of meeting similar fate if they oppose. The local police and administration is also insensitive and biased. In this murder case all accused got acquitted for lack of evidence within a period of six months.[3]
- **Fudging EIA reports**—All the EIA reports submitted by the industry for its expansions, and new projects have been misleading in land acquisition requirements and land-use pattern. For expansion project of JSPL (Rs. 2,060 crore). In 2005-06, the EIA report submitted used the data of 1991 census instead of 2001. The EIA misrepresented the actual number of villages stated that there were only 25 villages in a radius of 10 kms which was grossly inaccurate as per the SDM report highlighting 54 villages within the radius of 10 kms. While the total land already available to JSPL according to the application of Consent Form was 145 acres, the EIA Report stated it as 118 acres.[4]

- **Over Exploitation of Natural Resouces**—JSPL, unconcerned about the impact of its massive industrial activity on the local economy and natural resources, is extracting water from Kelu and Mahanadi rivers for its projects. None of the EIA report submitted by the industry so far has made any comment on the impact on riparian culture and local people lives of heavy water extraction. The industry is enjoying privileges of being rich and influential even over natural resources.
- **Environmental Clearance for the Expansion project**—Inspite constant protests on ground against the company and continuous submissions of complaints to the MoEF Expert Committee on Industrial Expansion and cases running against the industry in Environment Clearance Appellate Authority of MoEF highlighting several issues, the projects of this company have been getting environmental clearances. (*Infopack*, 2007)

Poorly Functioning Pollution Control Boards

Over the past few years, as the local agitations against Sponge Iron industries have strengthened, the Pollution Control Boards at the Regional and State-levels have been forced to take steps—like issuing closure notices—though for a temporary period, fining the companies involved or declaring certain areas as sensitive for further establishment of plants. The Orissa, Goa and West Bengal State Pollution Control Boards are ones who have taken some strong measures on this front.

There is also the issue of inability of the regional and state boards to monitor pollution, which is a general problem and not restricted to Sponge Iron factories. Once the consent orders are given, they are renewed inspite of poor compliance reports by officials during site visits. Many of the boards do not have the adequate equipment and staff required to collect data on air emissions.

1. "First of all they do not submit reports. Even if they do the actual concentration of SPM has never been brought out in the reports submitted," a senior official from Bellary Pollution Control Board stated adding "We all know that

the concentration level is very high and anyone can feel that. But to support industries, the reports are made to show that everything is fine with the air quality". The only testing laboratory where the ambient—air quality standard data can be tested is at Dharwad. According to some sources, the laboratory can be pushed to giving reports favourable to the industry. (*Source*: infochangeindia. org/200812267552/Environment/Features/Rifat Mumtaz/Dark-clouds-over-India 's-sponge-iron-industry. html)

2. After consistent pressure by the various movements against SIUs in the state in 2005, the Government of Orissa's Forest and Environment Department stipulated the regulations for the establishment of sponge, iron plants in the state of Orissa which involves no new sponge iron units establishments in six sensitive blocks. It also ordered strict following of pollution control norms but it failed to pressurize industries to do so.

   In June 2006, after a failure in getting any reasonable action on Bindal Sponge Iron unit in Angul District of Orissa, the local activists made a complaint against pollution from this SIU to the Supreme Court Monitoring Committee on Hazardous Wastes. The Committee which was on visit for mines inspection also visited Bindal sponge iron unit and made the following observations about the functioning of the plant —

   "The plant is improperly run without qualified people and is a classic example of how badly a large plant covering a vast area can be managed. The pollution from the plant has affected the health of twelve villages in the vicinity. The plant has no proper layout to segregate different handling areas. Water is flowing all over, with the contamination ultimately ending up in a pond which is only few feet away from the pristine Brahminy river. The entire plant air is full of fine dust which if inhaled can cause respiratory cancers. It is very much likely to affect the health of the workers as the Committee Members themselves could feel the severity of the pollution just as they entered into the premises. The plant, although it is only nine months old, looks like a few decades over. The Committee was informed that the plant had been given trial consent only considering the representations and protests from the nearby villagers. The unit does not have any boundary wall or even a sign board to indicate what it is operating and the type of operation being carried out, people involved, hazardous waste generated, etc. Considering the severity of the risks and hazards posed by the plant to the surroundings and the environment including the river, the Committee is of the opinion that the plant should be asked to stop the operation forthwith and it should be allowed to reopen only after it rectifies all the pitfalls including material handling, water collection, storage and treatment, emission control, EHS in the plant, etc. Everything is to be reorganized in terms of better material handling, EHS, water handling, etc. The plant should be allowed to restart only when an expert in sponge iron plants certifies that the necessary features have been installed and are running to his or her satisfaction." (*Source*: Supreme Court Monitoring Committee Report for Orissa, June, 2006)

In certain cases where communities or action groups have turned to the High Court in the face of inaction by the PCBs, they have got temporary relief in High Court orders—for instance in the case of Goa. However, as in most cases, it is difficult to find lawyers who would take up the issue or even judges who respond sensitively to the issue of pollution. The case of Goa has been shifted to the Supreme Court in 2008 due to failure of the pollution control board in pressurizing industries to comply with the orders passed for pollution control.

Key Issues Highlighted in the PIL in the Case of Jain Udyog-based in Sanguem Taluka of North Goa

1. Pollution of Kalay River, supplying drinking water to Opa Water Works, to Talukas of Ponda and Tiswadi including Panaji due to discharge of solid and liquid waste and even raw materials from the factory which is being erected on

an elevated land and arresting storm water in this Western Ghat region which receives 200+ inches rainfall, is impossible..

2. The project is being erected in an irrigated developed orchard by felling fully grown coconut and other trees. The investments made by the agriculturists in horticulture and other crops in the adjoining area are also sure to be destroyed, leading to loss of livelihood as well as health.
3. The site is located on the periphery of a Wild Life Sanctuary.
4. GPCB have reversed their own earlier decision of 21/11/2003, for unknown reasons, without considering the genuine objections of the affected citizens. Their Addendum dated 29/09/04 to the NOC ordering the company to locate their stocks 500 m away from Kalay river cannot be complied with, as they do not have any plot of land which is more than 250 m from Kalay River.
5. It is a fallacy to argue that the project should be allowed to come up first; and if it is found to be polluting, government authorities will take action, as is being suggested.
6. That even polluting projects should be allowed because they generate employment for local people is unacceptable. It is observed that local youth are neither keen nor found suitable by the promoters, to work in such hazardous industries. Almost entire labour employed for the existing units is migrated from other states, causing social, hygienic and other problems.
7. The company commenced construction of the project structures, even before the required clearances/NOC's were in place, and without complying with the laid down conditions.
8. GPCB's NOC was issued without assessing the impact of this and other similar projects in the vicinity, and without first verifying the company's tall claims regarding efficacy of the proposed pollution control equipments, or the feasibility/probability of compliance with its (GPCB's) own conditions. (*Infopack*, 2007)

The sponge iron industry has been identified under the 'red category' industry as being high in pollution impact. In a meeting of concerned citizens with the Member Secretary of the CPCB held in June 2008, the CPCB accepted the fact that the owners of sponge iron industry have been bad in controlling environmental pollution. It is worth-mentioning that inspite of possessing strong evidence and study reports done by various regional pollution control boards, the CPCB has not been able to tackle these issues considering restrictions of its official mandate.

State Government Support

It is the state governments who are today really providing a facilitative and pressure free environment for the Sponge Iron Industries, offering them every subsidy and incentive available. The newly formed states of Chhattisgarh and Jharkhand have evolved their industrial, water and power policies along the dictates of industrialists. In fact in Jharkhand the maximum number of SIUs mushroomed in Saraikela which was the constituency of ex-Chief Minister Arjun Singh Munda. In the Giridih district of the state SIUs are owned by local politicians. Having become disfavored in one location, the most common response of the sponge iron manufacturers is to relocate in another district and even state without too many queries from the governments on the reasons for relocation.

The Kundil Iron Ltd unit shifted from Goa to Londa for its SIU after facing stiff resistance from communities in Goa. It obtained its order for establishment from the Karnataka government in January 2005 followed by 21 acres of agricultural land close to the railway station.

The Karnataka State Pollution Control Board (KSPCB) had clearly stipulated that the company shall not commence construction before it obtained Environmental Clearance. But the company, in violation of these orders, commenced construction as reported in the press and seen physically on the ground in April 2005 without obtaining EC. The issue was vociferously raised by the press and NGOs. After this the KSPCB issued a show cause notice to the company in July 2005. In response, the company went to the Karnataka State Environment Clearance Committee (KSECC) and obtained the EC in July/August 05, after the show cause notice was issued by the KSPCB, who in the meanwhile had also filed a criminal case against the company for the violations. The

KSECC which gave the EC was chaired by the Principal Secretary, Karnataka who also is the chairman of the KSPCB. Thus he wore three hats at the same time and gave clearance—

(a) without considering company's violation of laid down rules, regulations and departmental directions, or despite all those violations,
(b) without calling for the opinion/comments of the forest department despite the SIU being classified a red category industry, highly hazardous in an ecologically sensitive area, and
(c) without consideration of SIU's pollution potential/effects on the unique and sensitive flora, fauna, water bodies, people, crops, natural resources like water, etc. and in violation of the national and state water policy priorities. (Infopack, 2007)

Subsidizing Sponge Iron Production Through Carbon Trade

While the players in the Sponge Iron industry are essentially domestic ones, ranging from the big wigs like Jindals, Tatas, Essar to the middle scale like Monnet and Vikram Ispat to the local MLAs and small time businessmen, what makes this industry truly global is presence in the International Carbon Trade market as a part of the Clean Development Mechanism (CDM).

The highly polluting coal-based SIUs sector seem to have found the CDM as a very profitable mechanism. The USP that is being used to sell carbon offsets by the Sponge Iron producing companies is WHR—or waste heat recovery. This basically means that the plant would use waste heat from the kilns and blast furnace gases to generate electricity, thus saving energy and utilizing waste gases and therefore reducing emissions.

The government of India is a signatory to the United Nations Convention on Climate Change (UNFCCC) and its legally binding Kyoto Protocol (since 2005) which is committed to addressing the issue of green house emissions mainly carbon emissions to address the issue of climate change. The CDM is a mechanism for the achievement of this objective which allows industrialised countries to meet their carbon emission reduction targets by paying for green house gas emission reduction in third world or developing countries.

The JSPL plant at Raigarh, Chhattisgarh, the largest producer of sponge iron in the world, with an annual turn over of Rs. 3000 crores has not one but four CDM projects approved by the National CDM Authority. Though, in reality, the official records and various complaints against the industry establish that it has been defaulter in controlling pollution.

The same logic goes for Tata, Ispat, OSIL and other large sponge iron producers who claim to be the first in using various technologies. As per the guidelines, companies are anyway expected to install Waste Heat Recover Boilers (WHRB) in their plants for this purpose—this has to be done irrespective of whether the project is a CDM project or not. As per a report in *Down To Earth (Gupta, R. 30th Nov. 2006)* "... any company would already be using the heat for power generation with or without CDM benefits if it has to make its venture financially viable. Moreover, as per law, gas with a temperature of 1,100℃ cannot be released into the atmosphere."

The sponge, iron industry all over India is highly notorious, with owners flouting all pollution control norms. The Ministry of Environment and Forests is well aware of the issues that have been repeatedly raised by the people in relation to violations in compliance of conditions, clearance procedures and public hearings by SIUs during establishment and even expansion. This makes it clear that the track record of the company as far as its environmental and social accountability goes is not a criteria for granting CDM approvals.

Some of the Sponge Iron company's who have received approvals for CDM projects include—the Orissa Sponge Iron Limited, Monnet Ispat, Tata Sponge Iron Limited, Jindal Steel and Power Limited, Ind Energy Private Limited, OCL India Limited, Balaji Sponge Iron Limited, Godawari Power, Ispat of the Hira group of industries (among the first to receive a CDM project).

In Conclusion: No Recourse on Environmental Destruction

The Sponge Iron Industry, with its clear backward and forward linkages of equally polluting activities—iron ore and coal mining on one hand and steel production on the other—is a classic case of how the poor continue to bear the brunt of large scale pollution of their living environment, while the polluters continue to flourish with abetment from every virtually dysfunctional government or regulatory authority in this country.

The struggles and movements against the highly polluting SIUs in various states like Chhattisgarh, Orissa, Jharkhand, West Bengal, Maharashtra, Karnataka Andhra Pradesh, Tamil Nadu and Gujarat—all demanding a right for a clean and safe environment for the poor—have the same stories to tell. Lack of political will and absence of corporate accountability dominate the scene. Economic viability of industry rules over viability of poor lives. In the case of sponge iron, it is the target of 10% growth in the steel sector which is the determining factor. Neither the government nor the corporates care about where the plants are sited, how they operate and who is bearing the costs.

It is an irony that India heads the Inter-Governmental Panel on Climate Change at the international level which is committed to minimizing environmental pollution. Quite contradictory to its international position, the Government of India has been amending various environmental regulations since 2004 favouring industries. The case of Sponge Iron industry pollution is only a tip of the iceberg. With slack environmental norms and no strict pollution control code, India is turning a blind eye towards the polluting sponge iron industry and allowing it to flourish.

Notes and References

1. *Kelo Pravah*, Raigarh, 6 January 2006.
2. Chhattisgarh State Electricity Regulatory Commission's Petition No. PL-3 of 2005.
3. www.cgnet.in
4. Gaps and points of contention for the Environment Impact Assessment (EIA) Report, JSPL Expansion Project, 2005.

References

Data and Analysis based on the primary survey, data collection and analysis collected in 2005-08, in the states of Chhattisgarh, Orissa, Jharkhand, Maharashtra, Goa, Karnataka and Andhra Pradesh.

Infopack on Sponge Iron Industries: Issues and Campaigns, National Centre for Advocacy Studies, 2007.

APPENDIX

National Seminar on 'Environment and Sustainable Development in India': A Report

Inaugural Session

The two day National Seminar by the CSESD, REU on *"Environment and Sustainable Development in India"* at the Kabi Janani Sarada Sabhakaksha in Rabindra Bharati University started at 10.30 A.M. on February 20, 2009, with the opening song by the students of the Rabindra Sangeet Dept. under the leadership of Dr. Indrani Ghosh, the Head of the Dept. and the welcome address by Dr. Raj Kumar Sen, Director CSESD and Professor and former Head, Economic Dept. After presentation of Flower Booguets to the dignitaries, Vice-Chancellor Professor Karuna Sindhu Das garlanded the portrait of Tagore and delivered his Inaugural speech. He welcomed this topic for discussion during the two days of the seminar. Talking on the subject he emphasized Rahindra Nath Tagore's stand on agriculture and environment. Quoting extensively from Gita, Puranas, Veda and Upanishadas, he formally declared the seminar inaugurated.

After the inauguration, Prof. V.R. Panchamukhi, the Chief Guest and former Chairman of the ICSSR and currently Managing Editor of the *Indian Economic Journal* discussed on "Globalization and Sustainabilities—Issues and Challenges". He started with the four pillars of sustainability—environmental, economic, socio-political and cultural in a holistic approach. He also analyzed the ancient Indian wisdom on environment which was distributed upon time, direction, conscience and mind or thoughts with the consideration of *prithvi* (land), *op* (water), tejas (fire), *vayu* (air) and *akasha* (space). After the setting up of Trilateral Commission in 1970s, the current phase of globalization brings into picture the concept of marketdriven, volatile capital flowing system. His suggestions for preserving environment in the era of globalization are to avoid jobless, ruthless, futureless, voiceless and rootless growth.

Prof. Deb Kumar Bose, the Guest of Honour and former Chairman, West Bengal Pollution Control Board and one of the Advisors, CSESD, was the next Speaker in this session. At the outset he refereed to Chapter I of the Generis of the Bible where God blessed Adam and Eve to be fruitful and multiply and dominate over all the earth and over every living thing that moved upon the earth. However, the results were disastrous and God repented his decision in Chapter 6, deciding to destroy man and the living things in nature excepting Noah, whom he held in trust. According to him, 23 per cent of plants are edible and 85 per cent human are using non-edible (toxic) plants for their daily consumption. As an example, he said that there were 10,000 varieties of wheat produced in China which was reduced to 22 only in 1997. He also observed that restriction in due for the industries like power, steel, cement and coal-based thermal power to save nature for sustainable future.

Prof. Binayak Rath, Vice-Chancellor of Utkal University and the Special Guest was the next to speak on 'Promotion of Mining Sector in India.' He discussed upon both the classical and neo-classical versions of ecology in the perspective of mining sector in India. According to him, India's reserves are 8% of world coal resources. Since 1994, when the Environment Impact Assessment (EIA) Rules started in India, holistic human development was a matter of concern. From a case study of Orissa hill areas, some projects were found in these zones to protect environment which includes Kalpataru, Samriddhi, Swawalambi, Nirmal Gram, Mamata, Kadambari and Neeradhara.

The next speaker in that session was the Special Guest Prof. M. Anisur Rahaman, Ex-vice Chancellor of Rajshahi University, Bangladesh who spoke on "Globalization and Sustainable Development." According to him, non-renewable resources are decreasing rapidly. Some major steps required for fair sharing of natural resources, avoidance of wasteful use of natural resources, sustainable food security and insects-related manure. Prof. Rahaman also made some suggestions in this regard. Those are: (i) biological diversity is required, (ii) homogenisation be stopped, (iii) large expanses of land, (iv) Management of fertilizers, (v) fundamental change in food supply is required, (vi) wild-life conservation and protection is required, and (vii) deforestation and overuse of under ground water be stopped.

This session was chaired by Dr. Sanat Kr. Ghosh, Dean, Faculty of Arts, Rabindra Bharati University. It came to an end with a vote of thanks by Dr. Amit Mukhejee, Finance Office, R.B.U. and Joint Director, CSESM.

TECHNICAL SESSION I

The Keynote address titled *"Sustainable Development and Environment : Lessons from Ancient Indian Wisdom"* was delivered by Dr. S.D. Chamola, Professor of Economics and Management (Rtd.), Chaudhary Charan Singh Haryana Agricultural University, Hisar.

The speaker divided the lecture into 3 main parts: I. Present Development Study, II. Lessons from Past, and III. Can we incorporate it in our practice?

He started with the view that present model of development is not harmonious and sustainable. For immediate gains we are becoming permanent loser. He said that the present model of industrialisation leads to violation of fundamental rights specially of farmers. Common property resources are being destroyed. The poor are getting deprived from livelihood. Corporatisation is creating growth based on faulty mechanism. Farmers and the poor are neglected and bypassed. Inequality in income and wealth is gradually increasing. Half of the population do not have any access to sanitation, proper education, etc.

With this erroneous process of development, environment is getting jeopardised. Economic exploitation and environmental destruction are going hand in hand. Environment is being destroyed for personal gain. There is also a huge disparity in the utilisation of resources by different nations.

Therefore, there is a need to rethink regarding our approach to sustainable development and environmental protection. In finding out in alternative model, many lessons can be learnt from Indian ancient civilisation where there was a harmonious relationship between man and environment.

There is a vast literature in ancient time like the Vedas, Upanishads, Ramayana, Mahabharata, Arthashastra, where environmental resources are worshipped and used on a sustainable basis. According to Vedas the earth is our mother and it is our duty to protect her from degradation. Upanishad says that the resources of this universe have to be consumed with restraint. In the epics also it has been said that resources should not be overexploited. Bhagabat Purana also concludes that resources should not be in the hands of the selfish fulers, rather they should be in the hands of noble people. Bhagabat Geeta says that if man protects nature, then nature will also protect man.

The Buddhists and Jains started revolution against animal sacrifice as rituals. They think that violence is the root cause of all economical as well as ecological destruction. Non-violence i.e., 'Ahinsa' is the only solution. Kautilya's Arthashastra holds that agricultural land, forest, animals, etc. should be protected for sustenance and this is the sciences of economics. Manusmriti similarly holds the view of maintaining harmony between man and nature.

Even in the recent times many communities like "Bisanoi" in Haryana and Rajasthan have given significant effort in protecting tree and wildlife. Mahatma Gandhi and his followers held the view that modern civilisation is endangering the environment for fulfilling the greed of human beings. Natural resources of countryside are being jeopardized. Rural and tribal communities are being deprived. Gandhi's philosophy was that trees and wildlife should be protected and he believed in the technique of non-violence. In very recent period Chipko Andolan, Narmada Bachao Andolan, etc. are some of the efforts put by man to protect the environment.

The next lecture was on "*Environment and Ecology in Kautilya's Arthasastra*" by Dr. Ratan Lal Basu, Former Teacher in Charge, Bhairab Ganguly College, Belgharia. It has been mentioned that Ecological concept of ancient India was different from present. Modern concept started in 1972 at Stockholm in the "United Nations Conference on Human Environment." Growth and evolution of the species is called ecology and we have to preserve ecology for our own interest. Any kind of ecological imbalance originates from greed (*Aranyadevata* by Tagore). In ancient India, all the literature like the Vedas, Upanishads, Epics, Dharmashastras, Arthashastras and others prescribed measures for preservation of ecology and environment. Among all the texts Kautilya's Arthashastra is the most pragmatic in its approach. It has given specifications on avoiding man made as well as natural hazards. The Arthashastra has given specified rules for preservation of forests, pastures and mines, it has considered all aspects of houses and dwelling places (sewerage etc.), suggested some punitive measures, prescribed laws for protection of agricultures and biodiversity, avoiding fire hazards, flood hazards, famines, etc. It is king's duty to adopt all these measures. In essence, a holistic approach should be taken and ethics should be there in protecting the environment.

The next lecture was on "Environment Performance Index (EPI) and its relationship with macroeconomic indicators: A model approach" by Ms. Amrita Pal, (Department of Economics) and Dr. Jayanta Sinha, (Department of Zoology) of B.B. College, Asansol. Over the past few decades, there has been an increasing consciousness regarding environment. The link between economic growth and ecological system is the main essence of sustainable development. The term first came in IUCN World Conservation Strategy. It is basically fulfilling the need of present generation without making the future generation worse-off. So it is important to know the performance of a particular country in maintaining this sustainability. Environment Performance Index (EPI) is a datum driven, fact-based expirical approach. It aims at replicating the policy priority of environmental authority. EPI was led by the Environmental Sustainability Index (ESI) (1999-3005). There was 7 macroeconomic indicators in EPI viz. GDP per capita (PPP US$), population (mn), life expectancy at birth (years), percentage of urban population, public expenditure on health (% of GDP) and CO_2 emission (total per year). EPI 2006 has been calculated for 133 countries, but here only 25 countries have been taken. Multiple regression models have been used to find out the mathematical relationship between EPI and macroeconomic indicators mentioned earlier. Data have been collected from Human Development Reports and EPI reports. Positive linear relationship is obtained between HDI and EPI (closest), GDP and EPI, life expectancy at birth and EPI, and CO_2 and EPI (lowest). Population and EPI are negatively related and other two have a quadratic relation with EPI. Here P values are significant but individuals are not. In conclusion the speaker said that for sustainability desired changes should come in with continuity.

Prof. Rabindranath Bhattacharya suggested since HDI and life expectancy at birth are related, one should not to take GDP per capita and life expectancy but to take HDI only.

Technical Session II

The session on sub-theme II was chaired by Prof. Rabindranath Bhattacharya, visiting Professor, Rabindra Bharati University. There were eight papers of which four are presented and one paper from sub-theme III was also included in this session.

The Key Note paper was presented by Prof. Kausik Gupta of Rabindra Bharati University on "Sustainable Management of Renewable Resources with special reference to India." In a lucid presentation, Prof. Gupta pointed out that development activities have strong impacts on environment. Both the Human Development Reports and the Millennium Development Goals focus on poverty, hunger and environmental sustainability. He divided his analysis in three parts – land management, fishery management and fresh water resources management. The study takes into account of various factors related to fresh water pollution and attempt to relate sustainable management of fresh water resources with sustainable management of "watershed plus". This paper also suggested some strategies for land regeneration in the context of a developing economy like India. Regarding sustainable management of forestry the paper also found that India is experimenting with diverse programmes for protection, regeneration and biomass production in forests.

Next paper was presented by Dr. Lekha Mukhopadhyay of center for rural and cryogenic Technology on "What causes variation in Efficiency of Labour for collection of Common Property Resources—Agro-Economic Conditions and/or conditions of Resource Base" which was a joint paper with Suranjana Joarder of Jogamaya Devi College. This paper mainly highlighted on the relationship between equity, efficiency and resources conditions as primary concerns in the new paradigm of the sustainable development. They have tried to find out the possible answer of variation of labour efficiency in forest-based community optionally using available forest land and labour time. Result shows that technical efficiency increases but allocative efficiency decreases with greater the livestock and household endows. And better is the agricultural production, technical and over all economic efficiencies of households if fodder collection increases.

The third paper in the session was presented by Shri Anirudha Ojha of Utkal University on "Poverty, Property Rights and Common Property Land Resources Management in the Tribal Belt of Orissa." This paper examines efficacy and implications of conferment of property rights for the effective management of common property land resources (CPLR) in the tribal economy of Orissa. The study makes some recommendations on the basis of the findings in respect of improving the sustainability of CPLR use in a tribal dominated area.

The next paper was a joint paper by Biswajit Ray of University of Calcutta and Dr. Rabindra Nath Bhattacharya of Rabindra Bharati University and presented by Sri Ray on "Collected Action and Performance of Institutions: Does Heterogeneity Matter in Joint Forest Management in India?" With a sample study of 383 households, the paper found that collective actions of the Forest Protection Committees (FPCs) are systematically associated with average forest stocks and increasing forest canopies, but negatively correlated with distances to the local forests among the forest attributors. The implications of this study are two-fold: (1) institutions should internalise group diversity to promote cost-effective collective actions through locally negotiated means, and (2) even if an institution has high collective action, it may not always be reflected in its performance.

The last paper in that session was also a joint paper by Mala Bhattacharyya and Rajrupa Mitra of Raja Peary Mohan College on "Ecotourism as an Environmentally Sustainable Development Process. A Case Study of Sonamukhi Forests (Bankura)". The study was based on a survey undertaken in the Sonamukhi Forest area and the findings reveal many inadequacies in the infrastructure like underdeveloped roads, transport and communication, shortage of power supply, water scarcity and other amenities. This paper mainly highlighted on the fact that on overall equitable, efficient and environmentally full-proof improvement of the region may be effected through ecotourism. This session was concluded with Chairman's observations and with a critical analysis of all the papers presented in this session.

The third technical session of the seminar began at 10.30 a.m. on 21st February 2009 (Saturday). It was chaired by Principal Fr.

Dr. Felix Raj and Professor Dhirendra Nath Konar, Co-chaired the session. At the outset the Chair made a very brief observation on the importance and relevance of the sub-theme. Environment and Sustainable Development in India—Regional/Sectoral and Empirical Studies.

Five papers were presented in this morning session. Anath Bandhu Mukhopadhyay presented the inaugural paper entitled, 'SEZ, Pollution and Sustainable Development'. In his paper Sri Mukhopadhyay made a scathing attach on the policy of creating Special Economic Zones in developing countries for rapid economic growth. He felt that this policy was nothing but a new Mantra to exploit the poor nations and cause irreparable damages to their ecosystem and environment by the developed countries. He stressed on the importance of living with the nature.

The second paper was presented by Sri Ashok Dasgupta. The title of his paper was, Quarrying the Relevance of the Term 'Indigenous People' in a three-level approach (with reference to Rajbansi Community of North Bengal). It is an anthropological study, which questioned the applicability of the term indigenous people to all the folk communities in general and the Rajbansi Community of West Bengal in particular. The attachment with land and historicity, role of indigenous knowledge system, and the effect of globalization are the three important criteria on which the decision should be based, he concluded.

The paper entitled, 'The Sponge Iron Industry Pollution : A Classic Example of Poor Environmental Governance was presented by Ms. Rifat Mumtaz. The Sponge Iron Industry in India is growing in strength though it is responsible for unprecedented levels of pollution. Thick black smoke, contaminated water, depleting vegetation, failing agricultural yields, premature death of domestic cattle and poor human health conditions are some of the impacts of this industry. The paper presenter deplored the fact that India heads the Inter-Governmental Panel on Climate Change at the international level, which is committed to minimizing environmental pollution. Quite contradictory to its international position, the GOI has been amending various environmental regulations since 2004 favouring industries. Thus with slack environmental norms and no strict pollution control code, India is turning a blind eye towards the polluting sponge iron industry and allowing it to flourish.

Sri Hari Govind Prakash presented his paper entitled, 'Environment, Sustainable Development and Slums'. The focus of the paper is urban poverty, which is manifested in the slums. The slums have posed several socio-economic and environmental problems for cities, which need remedial attention for sustainable economic development. He thus suggested formation of slum cooperatives for slum improvement and their upliftment.

The last paper of the session was presented by Sri Sushant Singh as the paper writers, Dr. V.D. Sharma and Brijesh Sharma were not present. The title of the paper was 'Environment and Sustainable Development in Indian Perspective'. The paper emphasizes the importance of the indigenous model of development as against the prevailing western model of development for the sake of sustainable development.

After the presentations were over, the Chairman of the session requested the audience to make observation or put questions, if any, on the papers presented in the session. Ms. Rifat Mumtaz, A.B. Mukhopadhyay, Dr. D.N. Konar, Dr. P.K. Pal, Dr. S.D. Chamola and Ms. Ratna Chakraborty participated in the lively discussion. While commenting on SEZ, Rifat Mumtaz observed that environmental impact assessment was a must for all types of SEZ. Dr. Konar wanted to know if Ms. Mumtaz used any statistical analysis in her paper to which she answered in the negative. Dr. P.K. Pal said that Ms. Mumtaz in her paper mentioned about forward and backward linkage effect of sponge iron industry. Did she have any data? She said that some data are available with the organization she was working with.

Fr. Felix Raj wondered if the third world countries are subjected to economic colonization since political colonization on is no longer possible. Constructive critical analysis on the happenings in India is needed to stay in the right path. He rounded up the session after thanking the co-chairman, the presenters of the five papers and the audience. In the last technical session, these papers were presented.

G. Satiskumar and S. Ramaswamy focused on new system including production, consumption and governance where three tier participating programme deals with better coordination of global environmental governance.

As far as Project Appraisal, Environmental Accounting and Sustainable Development is concerned, Pronab Nag and Anup

Kumar Sinha stressed on three ways of Projects Appraisal—Financial, Social and Economic, Pronab Nag discussed on Environmental Accounting which measures five environmental problems (UNCTAD) as depletion of non-renewable resources, depletion of ozone layer, global warming and waste disposal.

The last speaker was Dr. Sheela Dutta Ghatak, P.D. Women's College, Jalpaiguri. Her paper was on 'Relationship between Environment, Health and Education : Importance of Community Participation', she took up the case study of Jalpaiguri town and the problem of people living in slums. Her basic approach in the paper was to develop an equitable health care service that will meet the need of people living there using appropriate technologies to develop health information and awareness throughout the society.

The session came to an end after brief deliberations on the papers. Dr. Amit Mukherjee, Joint Director of the Centre raised the vote of thanks and the two-day national seminar came to a close.

INDEX